PHYSIOLOGY

MEDICAL EXAMINATION REVIEW

PHYSIOLOGY

Eighth Edition

700 Multiple-Choice Questions
with Explanatory Answers

Kalman Greenspan, PhD, FACC

Professor of Medicine and Physiology
Indiana University School of Medicine
Indianapolis, Indiana

Chief, Section of Physiology
Chief, Section of Emergency Medicine
Terre Haute Center for Medical Education
Terre Haute, Indiana

MEDICAL EXAMINATION PUBLISHING COMPANY

No responsibility is assumed by the publisher for any injury and/or damage to persons or property as a matter of products liability, negligence, or otherwise, or from any use or operation of any methods, products, instructions, or ideas contained in the material herein. No suggested test or procedure should be carried out unless, in the reader's judgment, its risk is justified. Because the drugs specified within this book may not have specific approval by the Food and Drug Administration in regard to the indications and dosages that are recommended by the authors, we recommend that independent verification of diagnoses should be made. Discussions, views, and recommendations as to medical procedures, choice of drugs, and drug dosages are the responsibility of the authors.

Medical Examination Publishing Company
A Division of Elsevier Science Publishing Co., Inc.
655 Avenue of the Americas, New York, New York 10010

© 1990 by Elsevier Science Publishing Co., Inc.

This book has been registered with the Copyright Clearance Center, Inc. For further information, please contact the Copyright Clearance Center, Inc., Salem, Massachusetts.

Library of Congress Cataloging-in-Publication Data

Medical examination review.
    Includes various editions of some volumes.
    Published 1960 - 1980 as : Medical examination review book.
    Includes bibliographical references.

    1. Medicine—Examinations, questions, etc.
RC58.M4   610'.76—dc19   61–66847   AACR 2   MARC

ISBN 0-444-01533-7

Current printing (last digit):
10 9 8 7 6 5 4 3 2 1

Manufactured in the United States of America

Contents

Preface

This eighth edition of *Medical Examination Review: Physiology* has been substantially revised and updated to keep in step with current trends in medical education and the continuing expansion of scientific knowledge. It is designed to help you prepare for course examinations, National Boards Part I, FLEX, and FMGEMS.

The range of subjects included in this book is based on the content outline of the National Board of Medical Examiners, which develops the question pool for the tests mentioned above, and reflects the scope and depth of what is taught in medical schools today. The questions themselves are organized in broad categories to give you a representative sampling of the material covered in course work, while helping you define those general areas to which you need to devote attention. For your convenience in selective study, the answers, with commentary and references, follow each section of questions.

Using this book, you may identify areas of strength and weakness in your own command of the subject. Specific references to widely used textbooks allow you to return to the authoritative source for further study. This book supplements the answers with brief explanations intended to prompt you to think about the choices — correct and incorrect — to put the answers in broadened perspective, and to add to your fund of knowledge. A complete bibliography appears at the end of the book. The questions and answers, taken together, emphasize problem solving and application of underlying principles as well as retention of factual knowledge.

Acknowledgment

The author gratefully acknowledges the assistance of Mark A. Tucker, BS, MS in the preparation of the manuscript for this edition.

1 Overview of Physiology

DIRECTIONS (Questions 1–13): Each of the questions or incomplete statements below is followed by five suggested answers or completions. Select the **one** that is **best** in each case.

1. Under normal external temperature conditions, the most important system controlling water excretion or loss is
   A. skin
   B. lungs
   C. kidneys
   D. gastrointestinal tract
   E. body hair

2. Salivary secretion
   A. is only affected by parasympathetic innervation
   B. contains an enzyme that must be stored in zymogen granules before it is secreted
   C. is stimulated by sympathetic innervation
   D. contains the enzyme salivary amylase
   E. is a passive process and does not involve any enzymatic processes

3. Which of the following is most related to the fact that, in the steady state, cell membranes are relatively more permeable to potassium than to sodium?
   A. Cells possess an outside-negative transmembrane resting potential
   B. Cells possess an inside-negative transmembrane resting potential
   C. Cells change volume in the presence of somatic pressure gradients
   D. Normal cells show a progressive diminution in the internal potassium concentration in the steady state
   E. Cells do not possess active ion transport systems

4. In negative water balance
   A. the cells and extracellular compartment are hydrated
   B. only the cells are hydrated
   C. the intracellular compartment is hydrated but the extracellular compartment is dehydrated
   D. the total body water content is reduced
   E. the cells and extracellular compartment are dehydrated

5. A sample of human erythrocytes is placed into a solution of plasma that has been made hyperosmotic with the addition of urea, a substance that is permeable to the red blood cell (RBC) membrane. Which of the following best describes the results?
   A. The cells would shrink transiently and tend to return their normal volume
   B. The cells would swell and hemolyze
   C. The cells would shrink and then swell and hemolyze
   D. No change in the cell volume would be observed at any time
   E. The cells would undergo rapid mitosis

6. Which of the following is NOT associated with orthostatic hypotension?
   A. Increased heart rate
   B. Constriction of visceral and skeletal muscle arterioles
   C. Increased sympathetic activity
   D. Increased respiratory depth
   E. Venoconstriction

7. Sebaceous glands
   A. arise independent of the follicular canal
   B. secrete sebum from a single layer of basal cells that line their lumen
   C. increase their activity under the influence of androgens
   D. secrete sebum only in response to adrenergic excitation
   E. are important for temperature control

8. The primary stimulus for increasing heat loss to maintain thermal homeostasis arises from
   A. temperature receptors in the skin
   B. cold receptors in the skin
   C. special thermal receptors in the veins
   D. special thermal receptors in the arteries
   E. the anterior hypothalamus following elevated temperatures above a specific value

9. Which of the following best describes the process of diffusion?
   A. A "downhill" process that is passive in nature (no metabolic energy is necessary)
   B. An "uphill" process that is active in nature requiring a metabolic energy input
   C. A driven process that is purely dependent upon the presence of a true driving force such as a pressure or voltage gradient
   D. A process by which only large molecular weight proteins may move across biologic membranes
   E. Requires metabolic energy

10. In Figure 1 assume that side A is the capillary, side B the interstitial fluid, and the line separating compartments A and B the vessel wall. In such a situation, which of the following statements is INCORRECT?

   A. The protein in the capillaries is relatively impermeable and will stay in the vessel

   B. The H_2O in side B will tend to diffuse into side A due to the plasma oncotic colloid osmotic pressure

   C. Any pathologic situation that causes the vascular protein to enter side B would result in the clinical condition of edema

   D. The capillary hydrostatic pressure would oppose H_2O filtration from side B to side A

   E. None of the above are correct, because in all living organisms the capillary proteins are freely diffusible and hence would come to osmotic equilibrium

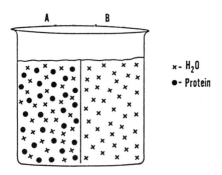

A   B

x – H_2O
● – Protein

Figure 1.

11. The best evidence that the plasma clearance of inulin (C_{in}) measures glomerular filtration rate in man is

   A. at high urine flows the clearance of urea approaches C_{in}

   B. it is found in the urine of aglomerular kidneys

   C. when phlorizin is given glucose and creatinine have the same clearance as inulin

   D. when the tubules are overloaded with para-aminohippurate (PAH) the clearance of PAH approaches C_{in}

   E. at low urine flows the clearance of glucose approaches C_{in}

12. Heparin prevents clotting primarily because it
    A. dissolves fibrinogen
    B. blocks conversion of prothrombin to thrombin
    C. inactivates thromboplastin
    D. chelates calcium
    E. blocks the action of thrombin *Heparin binds Antithrom III*

13. In a patient with diabetes mellitus who has a high blood glucose level and a filtered load of glucose above tubular transport maximum (T_m) level
    A. the polydipsia leads to plasma dilution, decreased plasma osmolality, and hyponatremia
    B. retention of sodium in exchange for potassium leads to a hypokalemic, metabolic acidosis
    C. the osmotic diuresis leads to a hyponatremic dehydration due to loss of sodium in excess of water
    D. there is base (sodium) conservation with reabsorption of bicarbonate in preference to chloride and resultant metabolic alkalosis
    E. there is metabolic acidosis that may be complicated by potassium retention since hydrogen is selectively secreted by the distal nephron in exchange for sodium

---

DIRECTIONS (Questions 14-32): Each group of questions below consists of a set of lettered components, followed by a list of numbered words or phrases. For **each** numbered word or phrase, select the **one** lettered component that is most closely associated with it. Each lettered component may be selected once, more than once, or not at all.

Questions 14-18 (Figure 2):
   A. Phase 4
   B. Phase 2
   C. Phase 3
   D. Phase 0
   E. Phase 1

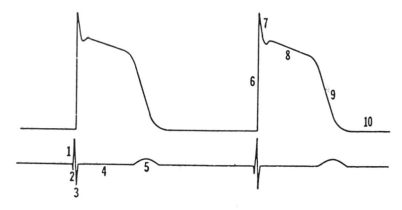

Figure 2. Recordings of the intracellular action potentials (top trace) and simultaneous electrographic tracings (bottom trace) from ventricular contractile tissue.

In Figure 2,

14. Deflection #6 is

15. Deflection #7 is

16. Deflection #8 is

17. Deflection #9 is

18. Deflection #10 is

Questions 19–25 (Figure 3):
    A. K^+ efflux
    B. Cl^- influx
    C. Fast Na^+ influx
    D. Slow Ca^{2+} influx
    E. Slow Ca^{2+} influx and simultaneous K^+ efflux

In Figure 3, the ionic movement responsible for current #1 is

19. current #1 is

20. current #2 is

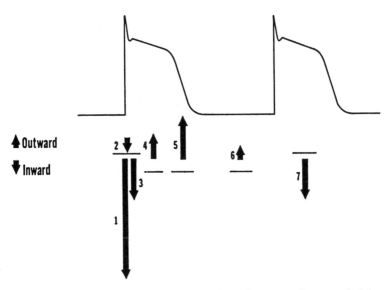

Figure 3. Arrows indicate the direction of current flow carried by ionic movement.

21. current #4 is A

22. current #5 is A

23. current #7 is D

24. current #3 is B

25. current #6 is A

Questions 26 and 27 (Figure 4):
    A. Curve A
    B. Curve B

In Figure 4, the curve which represents the relationship between pressure and flow for

26. an atherosclerotic (rigid) artery is A

27. a normal distensible artery is B

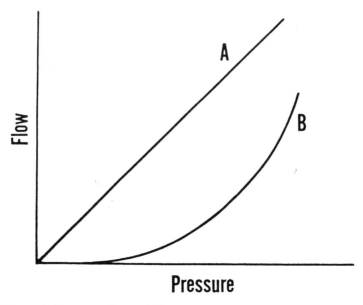

Pressure

Figure 4. Pressure-flow relationship of the blood in two different vessels.

Questions 28–32 (Figure 5):
- **A.** Site A
- **B.** Site B
- **C.** Site C
- **D.** Site D

In Figure 5, the site(s) at which there occurs the greatest or highest

28. urea concentration is

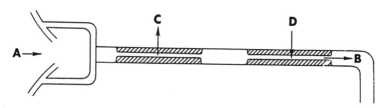

Figure 5. Schematic representation of a normal renal tubule.

29. K⁺ reabsorption is

30. K⁺ secretion is

31. creatinine secretion is

32. activity of antidiuretic hormone (ADH) is

DIRECTIONS (Questions 33–42): Each group of questions below consists of a diagram with lettered components, followed by a list of numbered words or phrases. For **each** numbered word or phrase, select the **one** lettered component that is most closely associated with it. Each lettered component may be selected once, more than once, or not at all.

Questions 33–35 (Figure 6):

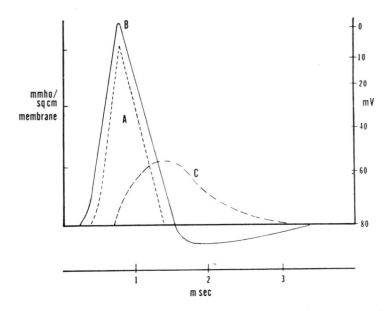

Figure 6. Relationship between membrane conductance and membrane voltage alterations as a function of time following a depolarization.

33. Action potential (voltage alterations)

34. Conductance changes exhibited by sodium (Na⁺) A

35. Conductance changes exhibited by potassium (K⁺) C

Questions 36–42 (Figure 7):

36. Local response

37. Negative afterpotential

38. Overshoot C

39. Spike potential B

40. Positive afterpotential

41. Level of potential at which the depolarizing process becomes self-generative (threshold potential)

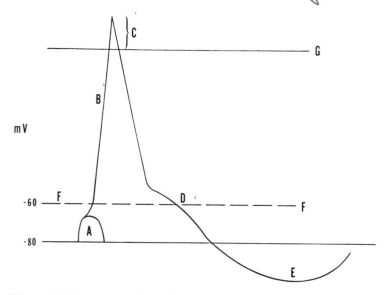

Figure 7. Representation of a nerve action potential.

42. Level at which no voltage gradient is present across the membrane

---

DIRECTIONS (Questions 43–87): For each of the questions or incomplete statements below, **one** or **more** of the answers or completions given is correct. Select

    A if only *1, 2, and 3* are correct
    B if only *1 and 3* are correct
    C if only *2 and 4* are correct
    D if only *4* is correct
    E if all are correct

---

43. Which of the following can be said with regard to the esophagus?
    1. Conduction of food through the esophagus is controlled by vagal reflexes
    2. The upper third of the esophagus contains only smooth muscle
    3. The esophagus contains both smooth muscle and skeletal muscle
    4. Conduction of food through the esophagus is controlled primarily by enteric reflexes originating in the esophagus

44. Which of the following may increase the force-generating capacity of a muscle?
    1. Increasing the diameter of a muscle fiber without an increase in fiber length
    2. Increasing the length of a muscle fiber without an increase in diameter
    3. Hyperplasia
    4. Parallel rather than pennate arrangement of cells

		Directions Summarized		
A	**B**	**C**	**D**	**E**
1,2,3	1,3	2,4	4	All are
only	only	only	only	correct

45. All of the following are characteristics of muscle tissue except:

   1. classified anatomically and functionally
   2. chemomechanical transduction
   3. individual cells of voluntary muscle tissue are arranged in parallel and function independently
   4. individual cells of involuntary muscle tissue are arranged in parallel and cannot function independently

46. Local spinal circuitry may be influenced by axons descending from the brain. The specific effect of this influence depends on
   1. discharge properties of the descending axon
   2. location or distribution of terminal axons
   3. physiologic inhibitory effect of the axon
   4. physiologic excitatory effect of the axon

47. Body fluid pH depends primarily on the renal control of
   1. reabsorption of sodium dibasic phosphate and secretion of bicarbonate
   2. reabsorption of sodium in exchange for potassium
   3. secretion of hydrogen ion in exchange for potassium
   4. secretion of hydrogen ion in exchange for sodium

48. The function(s) of the gallbladder include

   1. equalization of pressure in bile duct system
   2. reduction of alkalinity of bile pH
   3. concentration of bile fluid
   4. fluid reabsorption

49. The effects of spinal transection include all of the following except:

1. permanent anesthesia to body parts innervated by segments below transection
2. permanent loss of somatic reflexes below transection
3. permanent paralysis of voluntary musculature innervated by segments below transection
4. permanent loss of autonomic reflexes below transection

50. The lens of the eye

1. has a shorter focal length when relaxed
2. has a flatter lens conformity when relaxed
3. experiences less tension from support fibers when relaxed
4. is not relaxed during accommodation

51. Neuropeptides are found throughout the nervous system and play a central role in synaptic transmission. Some neuropeptides serve as neuromodulators whose functions include:

1. amplification of postsynaptic response
2. reduction of postsynaptic response
3. prolongation of postsynaptic response
4. increase-voltage of postsynaptic response

52. The primary cause(s) of heat stroke is(are)
1. excessive heat production
2. conductive heat gain
3. inability to lose heat by radiation
4. inability to sweat

53. Which of the following is(are) true regarding the chemical sensory mechanisms?

1. Limited to sensory mechanisms associated with taste
2. Include chemoreceptors located on the tongue and in the larynx and pharnyx
3. Limited to the olfactory sensory mechanisms
4. Of all the senses, the sense of smell is the least understood

Directions Summarized

A	B	C	D	E
1,2,3	1,3	2,4	4	All are
only	only	only	only	correct

54. Properties of mediated transport include:
    1. exhibit saturation kinetics
    2. transport across membrane at rates faster than predicted for molecules of similar size and lipid solubility
    3. exhibit competitive inhibition
    4. equilibration of substrate across cell membrane as in facilitated transport

55. In general, in receptive fields
    1. transducer-containing tissue changes its electrochemical properties in response to a stimulus
    2. a particular transducing tissue is relatively insensitive in that it is nonspecific for stimulus
    3. the transducer tissue response will perceive a specific stimulus from extreme forms of inappropriate stimuli as if the appropriate stimulus were there
    4. the ability to determine how much stimuli is present is due to the adaptation phenomenon

56. The sympathetic nervous system
    1. is predominantly excitatory
    2. is involved more frequently in generalized rather than discrete discharge
    3. is of little importance in visual accommodation of the lens
    4. may antagonize parasympathetic functions

57. Acclimatization to high altitude involves
    1. maintaining ventilation that is excessive with respect to CO_2 exchange
    2. producing an alkaline urine
    3. maintaining normal serum pH despite a low pCO_2
    4. increasing the oxygen-carrying capacity of arterial blood

58. Which of the following is(are) generally employed for long-term reduction of acid and pepsin secretion in a duodenal ulcer patient?
    1. Complete gastrectomy
    2. Daily administration of atropine in doses sufficient to block all action of acetylcholine
    3. Giving the patient plenty of aspirin in order to kill the pain, thereby reducing a stressful situation
    4. Gastric vagotomy plus pyloric antrectomy

59. The pyramidal tract arises from
    1. primary motor cortex
    2. somatic sensory areas of the cortex
    3. premotor cortex
    4. medulla

60. Stimulation of the myenteric plexus results in which of the following?
    1. Increased velocity of conduction along the gut wall
    2. Increased "tone" of the gut wall
    3. Increased rhythmic contractions
    4. Increased rate of rhythmic contractions of the gut wall

61. Properties of smooth muscle include:
    1. slow and sluggish contraction with the capability of sustained tonic contraction
    2. capability for rhythmic spontaneous contraction
    3. reciprocal innervation by sympathetic and parasympathetic systems
    4. insensitivity to mechanical, thermal, and chemical stimuli

62. With regard to gastric emptying
    1. the driving force of the emptying process is the pressure differential between the gastric and duodenal side of the pylorus
    2. it is influenced by duodenal pH
    3. it requires about 1 to 3 hours for a normal fixed meal
    4. it is facilitated by parasympathetic stimulation

Directions Summarized				
A	**B**	**C**	**D**	**E**
1,2,3	1,3	2,4	4	All are
only	only	only	only	correct

63. The normal fluidity of circulating blood is probably maintained by

1. plasmin
2. plasma antithromboplastin
3. plasma antithrombin
4. heparin

64. The volume of blood per total body weight

1. is constant for all members of a species; e.g., man
2. varies from about 50 to 100 mL/kg
3. is generally higher for women than for men
4. is greater in individuals with a greater total body weight

65. The oxygen-carrying capacity of whole blood is

1. about 20 vol%
2. largely determined by the plasma protein concentration
3. directly proportional to the hemoglobin content
4. increased in the presence of carbon dioxide

66. Receptors in joints are primarily associated with

1. somatosensation
2. thermosensation
3. chemosensation
4. proprioception

67. Physiologic levels of both growth hormone and thyroxine
1. increase the rate of protein synthesis
2. mobilize fat from adipose tissue
3. are necessary for normal growth and development
4. increase the peripheral uptake and utilization of glucose

68. Chief cells secrete

    1. zymogen granules
    2. intrinsic factor
    3. pepsinogen
    4. HCl

69. The collecting ducts are the site of final adjustment of urinary
    1. sodium
    2. potassium
    3. water
    4. pH

70. When using para-aminohippurate (PAH) to determine renal plasma flow (RPF) it is necessary to know which of the following values?
    1. Arterial concentration of PAH
    2. Urine concentration of PAH
    3. Urinary flow rate
    4. Urine pH

71. The initial step in the formation of urine is considered to be ultrafiltration at the glomeruli. The forces OPPOSING ultrafiltration include:

    1. colloid osmotic pressure of plasma protein
    2. glomerular capillary pressure
    3. hydrostatic pressure in Bowman's capsule
    4. crystalloid osmotic pressure of the final urine

72. During exercise increased O_2 to the active muscles depends on the following factors:
    1. skin temperature
    2. ventilation rate
    3. blood lactate concentration
    4. arteriovenous oxygen content difference

Directions Summarized				
A	**B**	**C**	**D**	**E**
1,2,3	1,3	2,4	4	All are
only	only	only	only	correct

73. The carotid sinus receptors in the adult human
    1. respond to relative hypoxia of the peripheral blood
    2. give rise to a depressor reflex if the blood pressure is too high
    3. are connected with cortical centers via the vagi
    4. continuously exert a tonic control in the vasomotor and cardioregulatory centers

74. Proteins produced by the platelets include:
    1. plasminogen
    2. fibrinogen
    3. prothrombin
    4. thromboplastin

75. Loss of heat occurs from the body core to the periphery and to the air by
    1. circulatory-assisted conductive and convective heat transport
    2. thermal radiation between body cells to the air
    3. the sweat glands
    4. convection from body cell to body cell and to the air

76. Initially, in the febrile or fever state
    1. the body loses its thermoregulatory capabilities
    2. body temperature continues to rise
    3. neurocontrol of peripheral vasculature is lost
    4. the hypothalamic set point temperature has been reset at a supernormal level

77. The contractile muscle of the heart is like skeletal muscle in that
    1. it is cross-striated
    2. the contractile elements are formed from myosin
    3. the force of contraction is increased by stretching (within physiologic limits)
    4. it does not possess automaticity

78. The stretch reflex will be evoked when certain receptors are affected. These are:
    1. annulospiral endings
    2. flower spray endings
    3. Golgi organ
    4. Meissner's corpuscles

79. Cellular membranes
    1. act as semipermeable barriers essential to maintaining an intracellular composition different from the outside environment
    2. play a critical role in the interactions between cells
    3. are responsible for forming compartments within cells
    4. vary in function, but structure and composition are the same among membranes of a single cell and from cell to cell

80. Prolonged hyperventilation may lead to respiratory alkalosis. This condition may be associated with
    1. increased renal reabsorption of bicarbonate
    2. decreased renal excretion of ammonium
    3. dissociation of acid buffers
    4. lowering of plasma bicarbonate concentration

81. A pure tone can be characterized by
    1. frequency and amplitude only
    2. amplitude and phase only
    3. phase only
    4. frequency, amplitude, and phase

Directions Summarized				
A	**B**	**C**	**D**	**E**
1,2,3 only	1,3 only	2,4 only	4 only	All are correct

82. Opioids

    1. are called morphinelike because they produce responses similar to morphine
    2. are produced in the central nervous system
    3. are produced in the pituitary gland
    4. seem to illicit complex behavioral changes including changes in mood and response to pain and stress

83. Which is not a function of the vestibular system?
    1. Senses forces in linear and rotational acceleration
    2. Stabilizes eye position
    3. Body balance and posture
    4. Spatial displacement of sound

84. During excess intake of fluids the extra volume of water is predominantly located in the
    1. intravascular compartment
    2. intracellular compartment
    3. interstitial compartment
    4. transvascular (GI tract and cerebrospinal fluid) compartment

85. Which of the following is a function of the cerebellum?
    1. Coordination and modulation of muscular activity
    2. Execution of body movement
    3. Integration of information concerning body position, muscle tension, and muscle length on a moment-to-moment basis
    4. Responsible for nonproprioceptive information

86. Swallowing
    1. is strictly under voluntary control
    2. is initiated voluntarily, but thereafter is under reflex control
    3. takes place in two stages: the voluntary stage and the pharyngeal stage
    4. inhibit respiration during the pharyngeal stage

87. To minimize sweating during exposure to the sun in a hot desert, one should
    1. wear as much black clothing as possible to reflect the rays
    2. removing clothing and walk slowly to increase convection currents
    3. walk clothed
    4. sit quietly and clothed (in white color and of texture that permits air flow)

Explanatory Answers

1. C. The kidneys are the supreme controllers of water excretion and account for over 60% of body water loss in the form of urine. But water loss also occurs via the lungs, skin, and the GI tract. (REF. 2, p. 1100)

2. D. The only enzyme secreted by the salivary glands is salivary amylase. (REF. 2, p. 1439 ff)

3. B. The greater permeability of a membrane to potassium will allow that membrane to develop a transmembrane potential that is largely a potassium diffusion potential. (REF. 2, p. 63 ff)

4. D. By definition, when total body water is reduced the body is said to be in negative water balance. (REF. 2, p. 1103 ff)

5. A. When human red cells are placed in plasma made hyperosmotic with urea the red cells will lose water until enough urea has moved into the red cell to reverse such loss. At this point, the cells will begin to return to normal size. (REF. 2, p. 868 ff)

6. D. The sudden drop in arterial pressure results in vagal withdrawal and an increase in heart rate. After a few minutes, sympathetic activity and circulating catecholamines serve to increase heart rate. However, the major effect of moving to a standing position is on the veins rather than the arterioles. Baroreceptor or chemoreceptor reflex elicits the abdominal compression reflex which helps translocate blood out of the abdominal vascular reservoirs; however, this reflex action of the abdominal muscles tends to limit rather than increase depth of respiration. (REF. 1, p. 248)

7. C. Testosterone stimulates both the growth and secretion of sebaceous glands. (REF. 1, p. 962)

8. E. The preoptic anterior hypothalamus is the major central area responsible for the sensing of body temperatures. (REF. 1, p. 853)

9. A. Diffusion is basically a passive process that can only occur down an electrochemical gradient. (REF. 2, p. 16 ff)

10. E. The distribution of fluid and ions between the vascular and interstitial compartments is controlled by the balance between hydrostatic and osmotic forces at the capillary level. The most important force is the capillary hydrostatic pressure which promotes filtration of H_2O and the water filtration is opposed by the plasma colloid osmotic pressure. Normally the vessel wall is impermeable to proteins. But a shift in membrane permeability to favor protein entry into the tissue could cause H_2O to enter the interstitial compartment and result in edema. (REF. 1, p. 386 ff)

11. C. The best evidence that plasma clearance of inulin (C_{in}) measures glomerular filtration rate in man is when phlorizin is given. Glucose and creatinine have the same clearance as inulin. Since phlorizin blocks sugar reabsorption, their clearance should then be identical with that of inulin. (REF. 2, p. 928 ff)

12. E. Heparin prevents clotting primarily because it inhibits the action of thrombin on fibrinogen and thereby prevents the conversion of fibrinogen to fibrin threads. It does this by increasing the efforts of antithrombin III. (REF. 1, p. 82)

13. E. The diabetic is constantly excreting high glucose amounts in the urine. The formation of ketones and decreased hydrogen for Na exchange yields a situation in which the patient is in a metabolic acidotic condition with loss of sodium, glucose, and large amounts of water. Since any Na^+ that is reabsorbed is exchanged for hydrogen, it leads to a K+ retention. (REF. 2, p. 1566 ff)

14. D. 15. E. 16. B. 17. C. 18. A. Transmembrane action potentials of the heart are descriptively labeled as phase 0 for the initial rapid depolarization (spike) including the overshoot. Phase 1 represents the initial rapid repolarization with phase 2 as the slow almost steady-state repolarization (plateau). The final repolarization curve is referred to as phase 3 which returns the membrane potential to its resting level (phase 4). (REF. 2, p. 782 ff)

19. C. 20. D 21. A. 22. A. 23. D. 24. B. 25. A. The cardiac action potential is a result of changes in membrane permeability to Na^+, Cl^-, Ca^{2+}, and K^+ going through "fast" or "slow" channels and controlled by "gates" which open or close in response to voltage changes. During phase 0 the Na^+ conductance is high with some influx of Ca^{2+}. Phase 1 is caused by an inward Cl^- current with a slight contribution by an outward K^+ current. The plateau or phase 2 is largely a continuous slow inward flow of Ca^{2+} with some contribution of a slow Na^+ inward. Also during phase 2 there is an anomalous rectification with K^+ moving in the outward direction. During phase 3 there is mainly a K^+ outward movement (through channels referred to as X_i). The resting level (phase 4) is maintained mainly by an outward K^+ movement. (REF. 2, p. 791 ff)

26. A. 27. B. The pressure-flow relationship for distensible blood vessels is not linear but curvilinear. In rigid vessels the flow is proportional to pressure in the range of 20–100 mm Hg mainly because the resistance to flow is constant. In a distensible vessel the resistance is changing $(1/R)$, which is due to a passive increase in vessel radius in response to increased pressure. As higher pressures are reached, the vessels passively dilate and thereby increase flow. So in a normal blood vessel the blood flow will increase in response to both an increased pressure and a decreased resistance, accounting for the curvilinear slope. (REF. 1, p. 211 ff)

28. B. The body normally forms about 25–30 g of urea each day and probably greater amounts in individuals on very high protein diets. This urea must be excreted in the urine, otherwise uremia will develop. (Normally the plasma concentration is 26 mg/100 mL, but in patients with renal insufficiency it can go up to 200 mg/100 mL). The rate of urea excretion is dependent upon glomerular filtration rate (GFR) and plasma concentration. At normal GFR about 60% of the filtered load is passed through the tubules, being higher or lower with changes in GFR. The lower collecting duct has the highest urea concentration, which is passively reabsorbed into the medullary interstitium and causes high concentration. The urea is then reabsorbed into the thin loop of Henle, so that it passes upward through these terminal renal portions several times before it is finally excreted. Despite the recirculation, none of the urea is actually reabsorbed into the blood but is eventually excreted into the urine in a very high concentration even though little water is excreted along with it, thereby the important role of urea in the execution of highly concentrated urine. (REF. 1, pp. 401, 417)

29. B. Potassium is transported through the tubules in almost identical fashion as sodium. About 65% of filtered K^+ is reabsorbed in the proximal tubules with 25% in the dilating segment of the distal tubules. By the end of the distal tubule only 10% of filtered K^+ is left. (REF. 1, p. 420 ff)

30. D. Considerable amounts of K^+ are secreted into the distal tubules. This K^+ is transported in the region of the tubule in a direction opposite to sodium (with the latter entering the cell) but not rigidly coupled with sodium. Either all or most K^+ movement is caused by the very negative electrical gradient created inside the cell when Na^+ is pumped out. The negativity pulls K^+ into the cell, which then diffuses out into the tubular lumen for excretion. (REF. 1, pp. 405, 420)

31. C. Creatinine in not readily absorbed but rather is secreted in the proximal tubule to actually increase the excretion of creatinine by approximately 20%. (REF. 1, p. 405)

32. E. The ADH secreted from the hypothalamic-posterior pituitary system results in a decrease in urine output. This hormone also causes an increase in H_2O reabsorption from the collecting ducts (and to a slight extent from the distal tubules as well). The urine that is excreted is thus highly concentrated. (REF. 1, p. 416 ff)

33. B. The action potential is characterized by an immediate voltage reduction and attains an amplitude of about 120 mV at its peak and is positive in polarity. This potential returns slowly to its original resting value after the occurrence of afterpotentials. (REF. 2, p. 57 ff)

34. A. The ability of Na^+ to cross the membrane is almost immediate after the membrane resistance changes. The unit of membrane resistance is in ohms/cm^2. (REF. 2, p. 57 ff)

35. C. After some delay (at peak of Na^+ conductance), the membrane reverses its permeability to favor K^+, and then is reflected by a K^+ conductance shape that demonstrates the delay and late onset. The K^+ conductance lasts until the membrane voltage is restored to its resting value. (REF. 2, p. 60 ff)

36. A. The slow charging of the membrane as the stimulus current is applied leads to membrane depolarization that is nonpropagated (the response is localized and does not spread very far). (REF. 2, p. 49 ff)

37. D. As the membrane potential repolarizes its rate, of repolarization is slowed for a few milliseconds prior to reaching its resting level. This period is called the negative afterpotential, and is believed to result from a K^+ accumulation just immediately outside the membrane. As a result, the concentration ratio of K^+ across the membrane is a little less than normal, preventing the immediate return of the potential to its steady-state value. (REF. 1, p. 112)

38. C. The overshoot represents the voltage above the zero level and is a result of excess Na^+ entering the cell. (REF. 2, p. 782 ff)

39. B. The spike is a term used to designate the sudden, rapid nerve and muscle depolarization resulting from the Na^+ influx (and to some extent Ca^{2+} influx). (REF. 1, p. 42)

40. E. When the excited membrane has repolarized to its resting level, the membrane potential may for a short period of time be a little more negative. This period of time is called the positive afterpotential. It is the potential which occurs toward the end of repolarization and is more negative by just a few millivolts. It lasts for a few milliseconds in duration and may be the result of excess efflux pumping of the Na^+. It is also a period of time during which the cell is in a less excitable state. (REF. 1, p. 109)

41. F. The threshold potential is that level of membrane voltage to which the membrane potential must be reduced before the Na^+ channels are all open and when the depolarization process becomes self-generative. (REF. 1, p. 113)

42. G. At a certain point during the depolarization process a sufficient amount of Na^+ ion has entered the cell to exactly neutralize the negative potential. The potential inside the cell is equal to the potential outside the cell and, in fact, no potential or voltage gradient exists across the membrane. (REF. 1, p. 108)

43. A. The esophagus is a muscular tube that conducts fluids and solids from the pharynx to the stomach during the third stage of swallowing. In humans, the upper third is skeletal muscle and during resting conditions is closed owing to the tonic contractions of the pharyngoesophageal sphincter. This prevents air from entering the esophagus during inspiration. Near the stomach region the esophageal muscles are of the smooth type and are not under conscious control. Initiation of the swallowing reflex is controlled by the vagus and the peristaltic conduction of food through the esophagus is primarily a continuation of contractions initiated in the pharynx. However, if primary peristalsis fails to move food through the esophagus, enteric reflexes originating in the esophagus take over. (REF. 1, p. 761 ff)

44. B. Increasing the diameter (hypertrophy) will increase the force generated by the muscle regardless of length. However, increasing the length by addition of sarcomeres without a change in diameter leaves the force-generating capacity unchanged. Also, the formation of more cells (hyperplasia) increases output. Finally, most cells are arranged at an angle to the axes of the muscle. This arrangement allows more fibers to attach to the tendon and increases total force-generating capacity. (REF. 2, p. 198 ff)

45. E. Muscle is responsible for chemomechanical transduction; conversion of chemical energy into mechanical energy. Muscle tissue is classified primarily two different ways: according to anatomy (smooth versus striated) and according to function (voluntary versus involuntary). Involuntary muscle cells are arranged in series as well as in parallel, forming a sheet of muscle. Consequently, the cells cannot function independently. (REF. 3, p. 315)

46. E. The physiologic effect of descending axons on local spinal circuitry may be inhibitory or excitatory. This is, of course, influenced by location or distribution of axons and their discharge properties. (REF. 3, p. 215)

47. D. The entire nephron is geared primarily for active transport involving $Na^+ - H^+$ exchange. Substitutes for Na^+ by other ions do occur, but these are secondary. Although H^+ secretion is determined by plasma K, it is the kidney which controls pH. (REF. 2, p. 1063 ff)

48. E. The major function is one of bile concentration. Other subsidiary functions include an ability to reduce bile alkalinity and the equalization of pressure within the bile system. The pressure equalization would be minimal or negligible if the gallbladder could not reabsorb fluid and reduce the bulk of the bile. (REF. 1, p. 781 ff)

49. B. Spinal transection results in immediate paralysis of all muscles innervated by spinal segments below the transection site. There is anesthesia in all body parts accordingly. However, although the somatic reflexes, such as muscle tone and the withdrawal response, along with the autonomic responses, may be abolished as well by transection, the loss is transient and is usually recovered, depending on the level of transection. (REF. 3, p. 199)

50. C. Relaxation refers principally to the circular musculature (ciliary body) around the lens to which tough support fibers (zonular fibers) are attached. In the relaxed state, tension from these fibers pulls the lens flat. Contraction of the ciliary bodies during accommodation relieves tension of the zonular fibers and the lens assumes a more spherical shape, which allows for closer focusing. (REF. 3, p. 98)

51. A. Though neuromodulators may prolong, shorten, reduce, and amplify the postsynaptic response, they do not cause substantive change in conduction or voltage. (REF. 3, p. 146)

52. D. Heat stroke is caused by failure of the sweating mechanisms. The reasons for this malfunction are not well understood. (REF. 1, pp. 859, 1017)

53. C. Olfaction (smell) and gustation (taste) are the only senses that can identify chemical stimuli. Chemoreceptors for taste can be found throughout the mouth, including the larynx and pharynx. Of all the senses, olfaction is the least understood, for at least two reasons. There are no odors analogous to sweet, sour, bitter, or salty taste. Second, sensory pathways themselves are difficult to characterize because of their diffuse nature. (REF. 3, p. 190)

54. E. All these properties are distinguishing features of mediated transport and are common to both active and facilitated transport. The distinguishing feature between these two is simply that active transport will "pump" against a concentration gradient whereas facilitated transport tends to equilibrate substances across the membrane. (REF. 3, pp. 16–17)

55. B. Transducer-containing tissue generally responds to a stimulus through electrochemical means. This response is relatively sensitive to one specific form of stimulus. Therefore, an inappropriate stimulus, if strong enough, will be perceived as the appropriate response. For example, if enough pressure is applied to the eye the retina will perceive the stimulus as flashes of light and not a mechanical stimulus. These tissues have developed not only the ability to determine how much stimuli is present but, through spatial distribution, can determine where the stimuli is located. However, slow adaptation sacrifices some precision in the intensity code in favor of an increased dynamic range. (REF. 3, p. 90)

56. E. The excitatory sympathetic nervous system is involved more frequently in generalized rather than discrete discharge. It is of little importance in visual accommodation of the lens, and antagonizes some parasympathetic functions. (REF. 2, p. 745 ff)

57. E. Acclimatization to high altitude involves all factors listed which consist of maintaining ventilation that is excessive with respect to CO_2 exchanges, producing an alkaline urine, maintaining normal serum pH despite a low pCO_2, and increasing the oxygen capacity of arterial blood. (REF. 2, p. 1038 ff)

58. E. The performance of a vagotomy will lessen the secretion of acid and pepsin by the stomach. The pyloric antrectomy is necessary to compensate for the decrease in the rate of gastric emptying. (REF. 2, p. 1438 ff)

59. A. Sixty percent of the pyramidal tract (also called the corticospinal tract) originates from the primary motor cortex, with 20% from the premotor cortex and 20% from the somatic sensory areas of the cortex. After leaving the cortex, the pyramidal tract passes through the brain stem, forming the pyramids of the medulla, then down into the corticospinal tracts. (REF. 1, p. 634 ff)

60. E. Stimulation of the myenteric plexus increases motor activity in the manner indicated in the question, i.e., increased tone, rhythm, rate, and velocity of contraction of the gut wall. (REF. 1, p. 757)

61. A. Smooth muscle is capable of sustained tonic contraction, although it is slow and sluggish. It is capable of rhythmic spontaneous contraction and it is reciprocally innervated by sympathetic and parasympathetic systems. (REF. 1, p. 146 ff)

62. E. Gastric emptying is influenced by duodenal pH and is facilitated by parasympathetic stimulation. Emptying takes about 1 to 3 hours for a normal meal. The driving force of the emptying process is the pressure differential between the gastric and the duodenal side of the pylorus. (REF. 1, p. 756 ff)

63. E. All the listed elements contribute to the maintenance of fluid blood. (REF. 1, pp. 58, 81 ff)

64. C. The volume of blood per total body weight will vary considerably from one individual to another, but it will be roughly proportional to body weight. (REF. 1, p. 276)

65. B. The capacity of blood to bind oxygen is fairly large and is proportional to the concentration of hemoglobin. (REF. 2, p. 1016 ff)

66. D. Proprioception is the sense of joint position. (REF. 1, p. 580)

67. A. Physiologic levels of growth hormone and thyroxine both increase the rate of protein synthesis and mobilize fat from adipose tissue, and they are necessary for normal growth and development. (REF. 1, pp. 887, 900)

68. B. The major secretion of the chief cells is pepsinogen in the form of zymogen granules (REF. 1, p. 774)

69. E. The collecting ducts of the kidney are final active areas for control of many substances found in the urine and play an important role in determining the pH of the urine. (REF. 1, p. 444 ff)

70. A. Effective RPF = urine concentration of PAH × urine volume/arterial concentration of PAH. (REF. 2, p. 928)

71. B. Both the colloid osmotic pressure of plasma protein and the hydrostatic pressure in Bowman's capsule tend to retard glomerular filtration. (REF. 2, p. 870 ff)

72. C. Increased ventilation and arteriovenous oxygen difference are two of the most important factors allowing increased oxygen supply to exercising muscles. (REF. 1, p. 1014 ff)

73. C. The carotid sinus receptors are primarily active in the control of blood pressure and are only a secondary influence on the control of blood pO_2. (REF. 1, p. 247 ff)

74. D. The platelets are the site for thromboplastin production, whereas prothrombin, plasminogen, and prothrombin are synthesized in the liver. (REF. 1, p. 790 ff)

75. B. Because of the volume of the human body the circulatory-assisted conductive and convective heat transport is essential for normal maintenance of body temperature. (REF. 2, p. 1584 ff)

76. D. The initial step in development of a febrile incident is the alteration of the set point of the hypothalamic temperature control center. (REF. 1, p. 853 ff)

77. E. All of the properties listed are shared by both muscle types. In particular, the true contractile cells of the ventricles of the heart do not, and cannot, spontaneously depolarize as the sinoatrial (SA) node of the heart. Nor do the ventricular cells become "automatic" under adverse conditions as can the Purkinje fibers of the heart. (REF. 1, pp. 120, 139, 150 ff)

78. A. The receptors that are needed to activate the stretch reflex are located within the muscle itself, and hence, it is the muscle receptor stimulation that induces the stretch reflex. (REF. 2, p. 538 ff)

79. A. Cellular membranes exhibit incredible diversity of structure, function, and composition. Though various membranes have some common features, membrane composition and structure differ not only from cell to cell but differ among membranes within a single cell. (REF. 3, p. 5)

80. C. Respiratory alkalosis is primarily a result of excess CO_2 that is blown off by the lungs, which results in a lower plasma bicarbonate concentration and concomitant increase in its reabsorption by the kidney. (REF. 1, pp. 406, 448)

81. D. Another parameter is needed to characterize sine waves, the mean level of the sine wave, "d.c." This is the average of peak to trough of the sinusoidal wave. However, in sound waves the d.c. level is not crucial as the d.c. refers to the ambient air pressure. (REF. 3, p. 158)

82. E. Many opioids have been found to be produced in widely scattered groups of cells within the body. The distribution of these cells include the cerebral cortex and spinal chord. They are also found in many synaptic terminals and axons. Their specific function, whether, analgesic or otherwise, is impossible to define for each of the opioids, but they seem to play an underlying role in the formation of complex behavior. (REF. 3, p. 147)

83. D. The principle function of the vestibular system is to sense forces due to acceleration, both linear and rotational and particularly the linear force of gravity. Actually, the vestibular system senses head acceleration. This sensory ability serves to stabilize eye position on a fixed point even though the head may be moving. It is essential for balance and posture. Even though the vestibular system is part of the inner ear, the vestibular pathways have nothing analogous to the sense of hearing. (REF. 3, p. 179)

84. D. Much of the extra water load (and up to water intoxication) moves into the cells of the blood and other tissues causing cellular hydration. (REF. 2, p. 1107 ff)

85. B. The cerebellum is involved with regulation of movement rather than its execution. It acts as a modulator of movement to smooth and coordinate movement. Of course this requires moment-to-moment input of information, all of which must be integrated and transmitted to descending motor pathways for ongoing activity. (REF. 3, p. 227)

86. C. Swallowing occurs primarily in three stages: the oral or voluntary, the pharyngeal, and the exophageal phase. The oral phase is strictly voluntary. However, after the bolus enters the pharyngeal stage involuntary reflexes take over. These reflexes are designed to propel the bolus towards the stomach at the same time preventing food from entering the trachea by inhibiting respiration and closing the epiglottis. This particular reflex takes place during the pharyngeal phase. (REF. 3, p. 662)

87. D. The best ways to minimize sweating are to avoid exercise, remain clothed, and attempt to reduce (if possible) direct exposure to the sun. (REF. 1, p. 849 ff)

2 Membrane, Neuromuscular, and Sensory Physiology

DIRECTIONS (Questions 88–117): For each of the questions or incomplete statements below, **one** or **more** of the answers or completions given is correct. Select

    A if only *1, 2, and 3* are correct
    B if only *1 and 3* are correct
    C if only *2 and 4* are correct
    D if only *4* is correct
    E if all are correct

88. The red reaction of the triple response is a skin reaction which is
1. due to a firm pressure with a pointed object
2. observed as a red line across the line of pressure
3. not dependent on nervous mechanism(s)
4. bounded by white areas

89. The monosynaptic reflex (as in the myotatic reflex) is
1. the result of spindle activation
2. facilitated by stimulation of gamma motor efferents to the homonymous muscle
3. excitatory to the homonymous muscle
4. inhibitory to antagonistic muscles

90. The diffuse character of pain from the gastrointestinal tract is probably due to the
    1. diffuse nature of pathways in the central nervous system
    2. absence of a visual correlate for the experience
    3. absence of segmental arrangement of pain fibers from the viscera
    4. sparsity of pain fibers compared to cutaneous area

91. Reflex stepping
    1. requires connections to the cerebellum
    2. is based primarily on unilateral cord reflexes
    3. does not occur if the plantar surface is prevented from touching the ground
    4. is probably modulated in the intact animal by group II fibers

92. Qualities of pain include:
    1. pricking
    2. aching
    3. burning
    4. referred

93. Myopia
    1. results when the image that is transmitted by the lens is in focus in front of the retina
    2. results when the image is behind the retina
    3. is corrected by a concave lens
    4. is corrected by a convex lens

94. The role of the parasympathetic nervous system in gastrointestinal motility is
    1. essential
    2. permissive
    3. unknown
    4. modulating

Directions Summarized

A	B	C	D	E
1,2,3 only	1,3 only	2,4 only	4 only	All are correct

95. When smooth muscle is stretched within physiologic limits
    1. the membrane repolarizes
    2. the tension that develops is because of elastic elements only
    3. syncytial conduction is enhanced
    4. the muscle contracts due to the depolarization

96. The distribution pattern of body hair is determined by
    1. androgens in men but not in women
    2. sex
    3. heredity in women, androgens in men
    4. sex and androgens

97. The saccule
    1. contains hair cells
    2. signals information about the head in space
    3. signals information about linear acceleration
    4. has a macula containing sensory cells that are directionally specific

98. Summation of muscle contraction can occur by
    1. recruitment of more motor units
    2. recruitment of additional muscle groups
    3. increased rate of stimulation of muscle fibers
    4. muscle hypertrophy

99. Pain receptors never are
    1. free nerve endings
    2. widespread in superficial layers of skin
    3. widespread in arterial walls
    4. encapsulated receptors

100. Glaucoma may be caused by
1. obstruction in the angle of the anterior chamber
2. obstruction at the pupil
3. obstruction in the canal of Schlemm
4. destruction of the ciliary secretory epithelium

101. Which of the following include sensations carried in the dorsal columns?
1. Joint sensation
2. Discriminative touch and vibration
3. Muscle spindle sensation
4. Heat

102. During photopic vision the
1. cones contain pigment that is sensitive to a specific color
2. rods are not stimulated
3. cones are responsible for most color distinction
4. visual acuity is lower than in scotopic vision

103. Pain control
1. can occur by activating the analgesia system
2. depends on an overstimulation of fibers
3. may occur by stimulation of the somatosensory endings in the skin over the area of pain
4. involves corticifugal signals

104. The clasp knife reflex
1. is the result of stimulation of a multisynaptic pathway and may protect the muscle from overload
2. is referred to as the inverse myotatic reflex
3. contributes to the smooth onset and termination of contraction
4. is caused by activation of the Golgi tendon organs and relaxes the muscle following contraction

Directions Summarized				
A	**B**	**C**	**D**	**E**
1,2,3	1,3	2,4	4	All are
only	only	only	only	correct

105. In the auditory system
1. there is tonotopic organization
2. there is multiplicity of tonotopic representation, each level comprising from two to five separate maps of the cochlea
3. extensive bilateral representation exists as a result of crossing and recrossing of fibers
4. the superior olive plays a major role in sound localization

106. Visual acuity is greatest in
1. an area that contains mostly rods
2. the fovea centralis
3. the lateral edges of the retina
4. an area that contains mostly cones

107. Chorea is
1. a constant, uncoordinated, random, uncontrolled flexing type of movement
2. the result of a lesion of the motor cortex
3. often seen in Huntington's disease
4. found only bilaterally

108. The frequency of action potentials in the nerve fiber (impulse rate)
1. is proportional to the amplitude of the receptor potential
2. is slower for receptor potentials at or just above the threshold and increases as the potential rises above threshold
3. have progressively shorter amplitudes with increase in frequency
4. is independent of the strength of the applied stimuli

109. Which of the following is true regarding physiologic optics?
1. Accommodation or changes in refractive power is accomplished by changes in the shape of the lens
2. Accommodation or changes in refractive power is accomplished by displacement of the lens relative to the retina
3. The major refractive power of the eye resides in the cornea
4. The major refractive power of the eye resides in the lens

110. The dermatome rule is used
1. to determine the extent of cutaneous tissue damage
2. to discern the slow pain response
3. clinically by physicians to determine levels of pain perception
4. to explain referred pain

111. The influx of sodium ions into a muscle fiber by the actions of acetylcholine
1. decreases membrane potentials (makes resting potential less negative)
2. increases membrane potential (makes resting potential more negative)
3. is ligand activated
4. is voltage activated

112. At the termination of an action potential, the potential sometimes fails to return to the resting level for another few milliseconds. Which of the following may be said regarding this phenomenon?
1. Termed the positive after potential
2. Results from buildup of sodium ions within the cell
3. Results from the excessive permeability of potassium ions
4. Most likely to occur after a series of rapidly repeated action potential

Directions Summarized

A	B	C	D	E
1,2,3 only	1,3 only	2,4 only	4 only	All are correct

113. The velocity of conduction in nerve fibers
    1. is roughly 0.5 m/s in unmyelinated fibers
    2. is as high as 100 m/s in myelinated fibers
    3. increases with the diameter in myelinated fibers
    4. increases with the diameter in unmyelinated fibers

114. Unilateral labyrinthectomy at the midpontine level causes
    1. spasticity of limb muscles on the side of the lesion
    2. a loss of all postural reflexes that have their centers in the medulla
    3. hyperextension of contralateral link
    4. no spontaneous nystagmus

115. In a nonactive nerve fiber in the body
    1. net efflux of sodium is greater than net influx of sodium
    2. passive diffusion of potassium outward is greater than passive diffusion of potassium inward
    3. passive diffusion of sodium inward equals the passive diffusion of potassium outward
    4. passive diffusion of sodium inward equals active transport of sodium outward

116. Depolarization of the neuromuscular junction is
    1. caused by acetylcholine (ACh)
    2. caused by nicotine
    3. blocked by curare
    4. blocked by atropine

117. The optic axis of the eye
    1. is the same as the visual axis
    2. passes through the nodal point
    3. is the axis along which the eye is directed
    4. is the axis of optical symmetry

DIRECTIONS (Questions 118–145): Each of the questions or incomplete statements below is followed by five suggested answers or completions. Select the **one** that is **best** in each case.

118. The "oscillatory" type of electrical activity recorded from visceral smooth muscle may be
    A. pacemaker potentials and initiated by Ca^{2+} influx
    B. pacemaker potentials and initiated by Cl^- influx
    C. pacemaker potentials and initiated by Na^+ influx
    D. local electronic change in potential not related to ionic movement
    E. pacemaker potential and initiated by K^+ influx

119. Sensory impulses arising in the periphery
    A. must all pass through the reticular formation
    B. never pass through the reticular formation
    C. may all pass through the reticular formation or may all bypass it depending on the modality
    D. pass in part through the reticular formation and in part bypass it
    E. ascend by unknown pathways to the cortex

120. Rapid adaptation of touch sensation is due to
    A. a decrease in firing rate despite continuous deformation of the receptor and/or return of the receptor to its original conformation despite continuous application of pressure
    B. maintenance of the distorted receptor conformation
    C. compensating mechanisms at the basal ganglia level
    D. failure of the cerebrum to detect a continuous sensory input
    E. failure of continuous sensory input to get through the reticular activating system

121. Visceral pain is most often due to
    A. electrical stimulation
    B. chemical stimulation
    C. stretch
    D. compression
    E. high body temperature

122. If you are placed in a chair and spun around from your left to right
    A. you will display left nystagmus
    B. you will display right nystagmus
    C. you will have your left semicircular canal maximally stimulated
    D. the pursuit phase will be to the left
    E. no nystagmus will occur

123. In certain pathologic conditions a sudden stretch of a skeletal muscle will result in sustained rhythmic contractions. This phenomenon is referred to as
    A. clonus
    B. hyperreflexia
    C. hyporeflexia
    D. spasticity
    E. clasp knife reflex

124. Receptors that mimic cholinergic receptors are called
    A. beta receptors
    B. serotinergic
    C. alpha receptors
    D. muscarinic
    E. dopaminergic

125. Which of the following is true with respect to the corticospinal pathway?
    A. It facilitates flexor motoneurons
    B. It is only a contralateral pathway
    C. It must be intact for a positive Babinski reflex
    D. It is an ipsilateral path at the level of the lower pons
    E. If it is severed at the level of the cerebral peduncles primarily extrapyramidal effects will appear

126. As we are all aware, when we lift a load with our hands the load is not moved with sudden, jerky motion but rather via a smooth-graded motion. This situation arises because the muscle
   A. asynchronously recruits its individual motor units
   B. contracts all or none in a tetanic fashion
   C. adjusts its elastic components to the load to absorb some of the kinetic energy of the contraction and there-fore to make it appear smooth
   D. neural impulse rate increases but number of units does not
   E. number of units for any muscle that fires is predeter-mined and constant

127. The auditory cortex is
   A. necessary for perception of temporal patterns of sound
   B. not necessary for understanding speech
   C. tonotopically organized
   D. represented by high frequencies in the anterior portion
   E. only found in the left hemisphere

128. "Stress relaxation" is a phenomenon seen in smooth muscle and allows organs lined by smooth muscle to act as reser-voirs (e.g., spleen, uterus). The capability of smooth muscle to demonstrate this phenomenon is caused by a property called
   A. tonus
   B. tetanus
   C. tone
   D. plasticity
   E. paresis

129. The CNS receives and processes information from the periphery. Which of the following choices BEST describes a method of sensory input?
    A. The magnitude of a sensory stimulus is received by the CNS and interpreted by changes in action potential conduction velocities
    B. The magnitude of a sensory stimulus is received by the CNS and interpreted by a change in action potential amplitude
    C. The magnitude of a sensory stimulus is received by the CNS and interpreted by a change in action potential frequencies
    D. The magnitude of a sensory stimulus is received by the CNS and interpreted by a change in awareness not involving action potentials
    E. The magnitude of a sensory stimulus is received by the CNS and interpreted according to the divergence of the stimulus during transmission

130. Hyperkinetic syndromes, such as chorea and athetosis, are usually associated with pathologic changes in
    A. the motor areas of the cerebral cortex
    B. the pathways for recurrent collateral inhibition in the spinal cord
    C. those portions of the reticular formation controlling the gamma innervation of muscle spindles
    D. the anterior hypothalamus
    E. the basal ganglia complex

131. The terminology used to denote deep sensation is
    A. proprioception
    B. exteroception
    C. chemoreception
    D. interoreception
    E. nociception

132. Neuromuscular transmission
    A. is caused by the release of ACh from the muscle side of the junction
    B. shows a permeability change to Na^+ and K^+ at the receptor site during the endplate potential (EPP)
    C. may be facilitated by curare in myasthenia gravis
    D. is blocked by curare because it competes with the Na^+ entry during the muscle action potential
    E. is solely an electronic function

133. Two-point discrimination is
    A. most effective on the hands
    B. most effective on the legs
    C. most effective on the face
    D. least effective on the face
    E. most effective on the big toe

134. The Golgi tendon organ
    A. decreases activity during strong muscle contraction
    B. stops activity during strong muscle contraction
    C. is a muscle-length receptor
    D. is a muscle-tension receptor
    E. is a motor effector organ

135. The visual cortex has
    A. simple cells
    B. complex cells
    C. hypocomplex cells
    D. distance cells
    E. bipolar cells

136. The muscles of the middle ear
    A. are always fully contracted
    B. protect the cochlea from overstimulation
    C. are necessary for auditory sensation
    D. are reflexly activated by loud sounds
    E. are vestigial in function

137. The action potential in a pacinian corpuscle starts at the
    A. end of the nonmyelinated fiber
    B. first node of Ranvier
    C. inside capsule layer
    D. second node of Ranvier
    E. outer capsule layer

138. The major cation directly involved in the isotonic contraction of skeletal muscle is
    A. Na^+
    B. Ca^{2+}
    C. K^+
    D. Mg^{2+}
    E. Cl^-

139. Contraction of muscles to hold the body upright against gravity depends largely on activity in the
    A. spinal cord
    B. cerebral cortex
    C. reticular formation
    D. hypothalamus
    E. basal ganglia

140. Alpha motoneurons
    A. supply the motor innervation for Golgi tendon organs
    A. supply the motor innervation for the smooth muscle fibers
    C. supply the motor innervation for muscle spindles
    D. are the final common pathway for the motor system
    E. supply the motor innervation for all glands

141. The generator potential of a receptor
    A. is fixed in amplitude and unrelated to stimulus intensity
    B. does not show degrees of adaptation
    C. is responsible for frequency-modulated nerve discharge
    D. is self-propagating and is nondecremental
    E. is entirely chemical in nature

142. All of the following are true in regard to the strength–duration curve EXCEPT
    A. it is a relationship between the duration of the stimulus and the amplitude of response
    B. it has a rheobase which is two times chronaxie
    C. cannot be characterized by its chronaxie
    D. it differs for different tissue; i.e., nerve and muscle
    E. it is necessary to determine the decremented characteristics of a nerve

143. When light rays come to a focus behind the retina the eye is said to be
    A. hypermetropic
    B. presbyopic
    C. astigmatic
    D. myopic
    E. emmetropic

144. For skeletal muscle one would expect an inverse relationship between
    A. muscle length and force of contraction
    B. load opposing contraction and velocity of contraction
    C. velocity of contraction (at low velocities) and efficiency of contraction
    D. muscle mass and cross-sectional area
    E. rest length and contracted length

145. A substance used experimentally to differentiate the end-plate potential from the muscle action potential is
    A. epinephrine
    B. histamine
    C. norepinephrine
    D. tubocurarine
    E. acetylcholine

DIRECTIONS (Questions 146–150): The following group of questions consists of a set of lettered components, followed by a list of numbered words or phrases. For **each** numbered word or phrase, select the **one** lettered component that is most closely associated with it. Each lettered component may be selected once, more than once, or not at all.

    A. Tonic neck reflexes
    B. Optical righting reaction
    C. Muscle spindle reflexes
    D. Tonic labyrinthine reflexes
    E. Crossed extensor

146. Is necessary for withdrawal reaction

147. Initiates contraction after rapid extension

148. When head turns to left, left foreleg extends

149. When neck is dorsiflexed, forelegs extend

150. After section of cranial nerve VIII the head can still be properly oriented

---

DIRECTIONS (Questions 151–169): Each group of questions below consists of a diagram with lettered components followed by a list of numbered words or phrases. For **each** numbered word or phrase, select the **one** lettered component that is most closely associated with it. Each lettered component may be selected once, more than once, or not at all.

Questions 151–154 (Figure 8):

151. Indicates the total muscle tension

152. Indicates the resting length

153. Indicates the tension developed by the passive elements

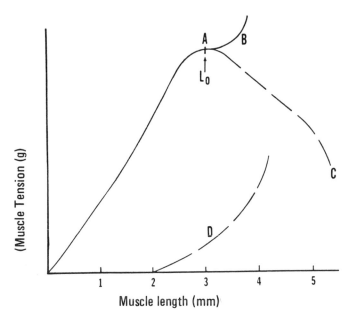

Figure 8.

154. Indicates the active tension developed by the contractile elements

Questions 155–158 (Figure 9):

155. Indicates the rheobase

156. Indicates the chronaxie

157. Indicates the utilization time

158. Indicates the strength–duration relationship

Question 159 (Figure 10):

159. The relationship between K^+ concentration and membrane voltage in living cells

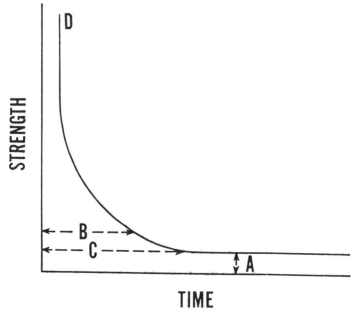

Figure 9.

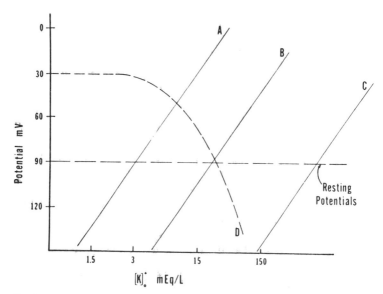

Figure 10.

Questions 160–169 (Figure 11):

160. The direction of K^+ flux during repolarization

161. The direction of Na^+ flux during depolarization

162. The direction of impulse propagation (orthograde)

163. The voltage gradient across the resting membrane

164. The voltage gradient across the membrane during activity

165. The fluid containing more K^+ (at rest)

166. The fluid containing more Cl^- (at rest)

167. The fluid containing more protein$^-$ (at rest)

168. The fluid containing more Na^+ (at rest)

169. Would indicate retrograde propagation

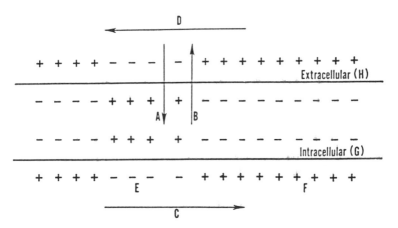

Figure 11.

Explanatory Answers

88. E. The red reaction to the triple response is the reaction to a firm stimulus of a pointed object over the skin area. The compress is a red boil (local red line due to venula dilation) with a pale area (on either side due to capillary constriction). This is not dependent upon nervous mechanisms since it can occur after autonomic nerve section and degeneration. (REF 2, p. 905)

89. E. The muscle spindle reflex originates in the muscle spindle whereby a type IA nerve fiber enters the dorsal root of the spinal cord and synapses directly with an anterior motoneuron, which transmits an appropriate reflex signal back to the same muscle. This results in its contraction. All the factors listed occur in this reflex. (REF. 1, p. 609)

90. D. The diffuse nature of visceral pain is established by the relatively light innervation of the area. (REF. 1, p. 598 ff)

91. B. If a well-heated spinal animal is held up from the table, and its legs allowed to fall downward, the stretch in the limbs occasionally elicits stepping reflexes that involve all four limbs. (REF. 1, p. 616)

92. A. Pain has been classified into three different types: pricking, burning, and aching pain. (REF. 1, p. 598)

93. B. Myopia is a condition in which the image of a distant object is formed not on the retina, but in front of it. In myopia, some of the excessive refractive power can be neutralized by placing in front of the eye, a concave spherical lens that will diverge rays. (REF. 1, p. 706)

94. D. The parasympathetic division releases acetylcholine and catecholamines which affect the frequency of spike discharge. This system does not initiate activity, but merely modifies or modulates an existing activity of the gastrointestinal system. (REF. 1, p. 757 ff)

95. D. Stretch is the most common physiologic stimulus for smooth muscle depolarization leading to contraction. (REF. 1, p. 143)

96. D. The genetic layout that will impart the characteristics of an individual plus the presence and exposure to androgens required for hair distribution determine the distribution of hair. (REF. 1, p. 961)

97. B. Located on the wall of both the utriculus and saccule is a small area slightly more than 2 mm in diameter called a macula. Each macula is a sensory area for detecting the orientation of the head with respect to the direction of gravitational pull or of other acceleratory forces. Each macula is covered by a gelatinous layer in which many small calcium carbonate crystals, called otoconia, are embedded. Also in the macula are thousands of hair cells, which project cilia into the gelatinous layer. Around the bases of the hair cells are intwined sensory axons of the vestibular nerve. (REF. 1, p. 621)

98. A. The stimulation of more motor units, increasing rate of stimulation, and recruitment of additional muscle groups will increase motive force. Hypertrophy associated with exercise will allow a greater total force. (REF. 1, p. 131)

99. D. Pain receptors are not encapsulated, are not easily adaptive, and respond to tissue damage, heat, and ischemia. (REF. 1 p. 593 ff)

100. A. Glaucoma is a disease of the eye in which intraocular pressure becomes pathologically high. In essentially all cases of glaucoma, the abnormally high pressure results from increased resistance to fluid outflow at the iridocorneal junction. The aqueous humor is produced by the ciliary secretory epithelium so its destruction should cause intraocular pressure to decrease. (REF. 1, p. 379 ff)

101. A. On entering the central nervous system, the thermal fibers travel in the lateral spinothalamic tract. But joint and muscle sensation and discriminatory touch are carried in the dorsal column. (REF. 1, p. 582 ff)

102. B. Color vision is mediated primarily by the cones which contain three pigments sensitive to blue, green, and red. (REF. 1, p. 716)

103. B. By stimulation of large sensory neural fibers the mechanoreceptor sensory afferents can suppress pain signals. Also the analgesia system can block pain at the initial entry point to the spinal cord. (REF. 1, p. 596)

104. E. In a spinal animal, the tendon reflexes that regulate tension are hyperactive and clonus occurs; sustained stretch reflexes are observed. This inverse myotatic reflex is a lengthening or clasp knife phenomenon and follows the four factors listed. (REF. 3, p. 209 ff)

105. D. Simple hearing functions such as threshold sensitivity, recognition of pure tones, frequency discrimination, and intensity discrimination are not severely impaired by rather large experimental lesions of central auditory structures. (REF. 1, p. 734 ff)

106. C. The fovea centralis in the center of the macula is best able to distinguish visual stimuli. This area is made up almost entirely of cones and requires higher light intensities for effective function. (REF. 1, p. 7116)

107. B. The cardinal criteria for the diagnosis of Huntington's disease are choreoathetoid hyperkinesis and hereditary disposition characterized by flicking movements which become more severe with time. (REF. 1, p. 629)

108. A. When a stimulus, such as pressure, is applied to a receptor, the receptor fires at a rate that is proportional to the applied pressure. Those impulses that result in action potentials with take-off potentials at or near the firing threshold generally have greater action potential amplitudes and are therefore lower in frequency. As the frequency increases the action potential amplitude decreases. (REF. 1, p. 574 ff)

109. A. The cornea is responsible for most of the refractory power of the eye (43 diopters); however, this is fixed. The lens has less power but can adjust (13 to 26 diopters). Accommodation is accomplished by changes in the shape of the lens, with some displacement of the lens relative to the retina. (REF. 3, p. 96)

110. D. Referred pain is the perception of visceral pain as emanating from a cutaneous area of healthy tissue. The explanation of this phenomenon is that the area in which the pain is felt is innervated by neurons from the same spinal segment innervating the affected organ. (REF. 3, p. 151)

111. B. Excitation of a skeletal muscle fiber by the action of acetylcholine results in a ligand-activated influx of sodium ions. This results in an increase in the positive direction of the local membrane potential. (REF. 1, p. 137)

112. D. Rapidly repeated action potential results in a buildup of potassium ions on the outside surface of the cell. This delays the return of the potential to resting level values. This is termed negative after-potential. (REF. 1, p. 112)

113. A. For unmyelinated fibers, the velocity increases with the square root of the fiber diameter and directly with diameter in myelinated fibers. (REF. 1, p. 114)

114. B. Unilateral labyrinthectomy results in limb rigidity in the side of the lesion and hyperextension and increased tendon reflexes in the contralateral limbs. (REF. 1, p. 620 ff)

115. C. A normal nerve at rest (nonactive) has a membrane potential of about -90 mV. This is maintained by the active transport system of the coupled Na^+-K^+ pump. Thus, the normal passive diffusion of Na^+ inward is counterbalanced by the active outward transport of Na^+. The normally greater outward diffusion of K^+ is balanced by the pumping of K^+ back into the cell. (REF. 1, p. 104)

116. A. The neuromuscular junction contains nicotinic cholinergic receptors. They are stimulated by nicotine and ACh and inhibited by curare. (REF. 1, p. 689)

117. C. The optic axis is different from the visual axis. Though they both pass through the nodal point the visual axis is the line along which the eye is directed while the optic axis includes a line that runs through the center of the cornea, pupil, and lens. (REF. 3, pp. 93–94)

118. A. The instability of the membrane, its tendency for spontaneous depolarization in visceral muscle, may be related to the fact that the smooth muscle membrane has more voltage-gated calcium channels than sodium channels. The flow of calcium into the interior of the fiber is responsible for the slow action potentials of smooth muscle. (REF. 1, p. 142)

119. D. It is a general rule of the nervous system that sensory impulses arising in a primary sensory neuron are distributed by way of collaterals, projection tracts, and secondary relays to widely separated parts of the nervous system. This is true of the reticular system as well, since sensory impulses partly pass through it and partly bypass it. (REF. 1, p. 580 ff)

120. A. The corpuscular structure of the pacinian corpuscle rapidly adapts to the deformation of the tissue because fluid within the capsule redistributes itself so that the pressure becomes essentially equal, thus reducing the receptor potential. (REF. 1, p. 575 ff)

121. C. Stretch due to distension is the most common cause of visceral pain. (REF. 1, p. 598 ff)

122. A. The ocular movements are in the same direction as the convection hydraulic current set up in the semicircular canals. The nystagmus consists of a slow (pursuit) and a fast (saccadic) movement, and it is named in accordance with the direction of the fast movement. The slow movement is toward the side stimulated. Thus, one would display left nystagmus. (REF. 1, p. 625 ff)

123. A. Clonus is the name used to describe regular rhythmic contractions of a skeletal muscle in situ under sustained stretch. (REF. 1, p. 612)

124. D. Acetylcholine activates at least two different types of receptors. These are called muscarinic and nicotinic receptors. (REF. 1, p. 689)

125. A. Flexor motoneurons are predominantly facilitated by the corticospinal pathway. (REF. 1, p. 613)

126. A. Special neurogenic mechanisms in the spinal cord normally increase both the impulse rate and the number of motor units firing at the same time. The tension exerted by the whole muscle is still continuous and nonjerky because the different motor units fire asynchronously. (REF. 1, p. 132)

127. C. Certain parts of the primary auditory cortex are known to respond to high frequencies and other parts to low frequencies. In monkeys, the posterior part of the supratemporal plane responds to high frequencies, while the anterior part responds to low frequencies. Thus, discrete tones are localized in discrete regions of the auditory cortex. (REF. 1, p. 741 ff)

128. D. A very important characteristic of smooth muscle is its ability to change length greatly without marked changes in tension. This results from fiber rearrangement due to smooth muscle plasticity. (REF. 1, p. 146)

129. C. The frequency of action potentials in the nerve fiber (impulse rate) is almost directly proportional to the amplitude of the receptor potential. (REF. 1, p. 574 ff)

130. E. Chorea is associated with degeneration of the caudate nucleus; athetosis is caused by lesions in the lenticular nucleus. (REF. 1, p. 329)

131. A. Events in the external and internal environments are first detected by receptors that can be categorized according to where the input is initiated: exteroceptors in the skin, proprioceptors in deep tissue (muscles, tendons, and joints). (REF. 1, p. 581 ff)

132. B. The amplitude of the action potential at the neuromuscular junction depends on permeability changes to Na^+ and K^+ at the receptor site during the EPP. (REF. 1, p. 136 ff)

133. A. Two-point discrimination consists of two needles pressed against the skin, and the subject determines whether he feels two points of stimulation or one point. On the tips of the fingers, a person can distinguish two separate points even when the needles are as close together as 2 mm. (REF. 1, p. 587)

134. E. The Golgi tendon organ detects tension applied to the tendon by muscle contraction. The signal from the tendon organ supposedly excites inhibitory interneurons and these in turn inhibit the alpha mononeurons to the respective muscle. (REF. 1, pp. 608, 612)

135. B. The majority of neurons in the striate cortex and all of them in extrastriate cortex possess receptive fields that are more elaborately organized than those of simple cells. Complex neurons are found in all cortical layers. They respond optimally to a properly oriented slit, edge, or dark bar, and for any one neuron certain spatial patterns are more effective than others. (REF. 1, pp. 653, 725 ff)

136. D. When loud sounds are transmitted through the ossicular system into the central nervous system a reflex occurs after a latent period of only 50 msec to cause contractions of both the stapedius and tensor tympani muscles. This attenuation reflex can reduce the intensity of sound transmission by as much as 30 to 40 dB. (REF. 1, p. 734)

137. B. Deformation of the capsule causes a sudden change in the membrane potential by increasing its permeability and allowing positively charged sodium ions to leak to the interior of the fiber. This change in local potential causes a local circuit of current flow that spreads along the nerve fiber to its myelinated portion. At the first node of Ranvier the local current flow initiates action potentials in the nerve fiber. (REF. 1, p. 575)

138. B. Evidence suggests that excitation–contraction coupling is effected simply by release of Ca^{2+} from the sarcoplasmic reticulum. The subsequent translocation of Ca^{2+} to troponin with consequent activation of actomyosin ATPase results in association of actin and myosin when the Ca^{2+} concentration has been raised sufficiently. (REF. 1, p. 127)

139. C. When a person or an animal is in a standing position continuous impulses are transmitted from the reticular formation into the spinal cord and then to extensor muscles to stiffen the limbs. (REF. 1, p. 619 ff)

140. D. The alpha motoneurons give rise to large type A alpha nerve fibers that innervate the skeletal muscles. They are the final common pathway for any motor response. (REF. 1, p. 606)

141. C. In all sensory receptors, the amplitude of the generator potential increases as the strength of the stimulus increases, but the additional response usually becomes progressively less as the strength of stimulus becomes great. (REF. 1, p. 574)

142. E. The least possible voltage at which a nerve will fire is called the rheobase, and the time required for this least voltage to stimulate the fiber is called the utilization time. If the voltage is increased to twice the rheobase voltage, the time to stimulate the fiber is called the chronaxie and is a means of expressing the excitability of different tissues. (REF. 1, p. 116)

143. A. When the axial length of an eye is too short relative to its focal length, the retina will intercept the bundle of rays from a distant object before it comes to a focus. A person with this kind of defect is said to be hypermetropic. (REF. 1, p. 705)

144. B. A decreasing velocity of a skeletal muscle seems to be caused mainly by the fact that a load on a contracting muscle is a reverse force that opposes the contractile force caused by muscle contraction. Therefore, the net force that is available to cause velocity of shortening is correspondingly reduced. (REF. 1, p. 126 ff)

145. D. The endplate potential and the action potential may be separated by utilizing d-tubocurarine, which inhibits and reduces the postsynaptic depolarizing action of ACh. (REF. 1, p. 137 ff)

146. E. Crossed extensor reflex facilitates support of the body by the contralateral side when a nociceptive withdrawal response is elicited. (REF. 1, p. 635)

147. C. When a muscle is stretched, muscle spindles are activated which reflexly causes contraction of the same muscle. This is the basis of the stretch reflex. (REF. 3, p. 629)

148. A. Input from tonic neck receptors has a major effect on tone in the forelimbs. (REF. 1, p. 613)

149. D. The tonically active receptors of the labyrinth affect fore-limb tone through the intermediate of neck muscle tone. (REF. 3, p. 271)

150. B. The visual system plays the most prominent role in right-ing reflexes. (REF. 1, p 609)

151. B. The total muscle tension represents the tension being contributed by both the active and passive elements and, therefore, will continue and increase as the length is increased. (REF. 2, p. 203 ff)

152. A. The resting length, or L_o, is the muscle length that is set in the body and represents the maximal overlap between the actin and myosin. It therefore will produce a drop in tension from that obtained at L_o. (REF. 2, p. 203 ff)

153. D. The tension developed by the passive elements is the result of stretching of these elements, which behave like stretching a rubber band. Such tension will therefore continue to increase as stretch increases until all elements are broken. (REF. 1, p. 125)

154. C. The actin-myosin components yield the true muscle ten-sion, and this tension is maximal at L_o. It may be obtained by subtracting the passive element tensions from the total muscle tension. (REF. 1, p. 125)

155. A. The rheobase is a voltage that is sufficient in strength to evoke a response when the duration of the stimulus is permitted to last for a long time, even infinitely if necessary. (REF. 1, p. 117)

156. B. Chronaxie is the time that it takes to evoke a membrane response when the strength of the stimulus is set at twice the rheobasic value. (REF. 1, p. 117)

157. C. When the stimulus is permitted to flow for a long period of time, even to infinity (to ascertain the rheobase current), there will be a response (at the rheobase strength) called the utilization time. (REF. 1, p. 119)

158. D. The slope of a strength–duration curve is exponential, and the parameters are inversely related. (REF. 1, p. 116)

159. A. At a normal resting potential of about -90 mV, the $[K_o]$ is about 3 mEq/L. Increasing the $[K_o]$ will depolarize the membrane. Such changes do not occur when the Na^+ and Cl^- concentrations are altered. (REF. 1, p. 102)

160. B. 161. A. 162. D. 163. F. 164. E. 165. G. 166. H. 167. G. 168. H. 169. C. An impulse is essentially a series of currents flowing in and out of the cell membrane. Action potentials leave the membrane ahead of the region of depolarization and act as a cathode (current flow into a cathodal electrode). The action potential current acts on the membrane ahead of it. At this time the membrane permeability is such that Na^+ can move inside while the area behind it is repolarizing and K^+ ion flux is outward. As thus defined, the impulse in the figure is propagating from right to left (i.e., arrow D). At rest, the voltage gradient is positive outside and negative inside the cell with the concentration of Na^+ and Cl^- greater with extracellular fluid, while K^+ and anion protein molecule concentrations are greater in the intracellular fluid. (REF. 1, p. 102 ff)

3 CNS Physiology

DIRECTIONS (Questions 170–238): For each of the questions or incomplete statements below, **one** or **more** of the answers or completions given is correct. Select

    A if only *1, 2, and 3* are correct
    B if only *1 and 3* are correct
    C if only *2 and 4* are correct
    D if only *4* is correct
    E if all are correct

170. Blood pressure and the blood supply to various organs are in part regulated by variation in the caliber of the small blood vessels and controlled by a medullary center under the influence of
    1. chemical changes in the blood
    2. blood pressure changes detected by baroreceptors widely distributed along the great vessels and in the viscera
    3. afferent impulses from visceral organs
    4. impulses from higher areas by which emotional activities are integrated

171. In regard to the "eating centers," they are
    1. found in the hypothalamus
    2. necessary for normal food intake
    3. antagonized by a satiety center
    4. served by glucose receptors in the cells of the hypothalamus

172. Presynaptic inhibition depends on
    1. reduced action potential amplitude
    2. reduced transmitter release
    3. reduced postsynaptic potential
    4. the inhibitory synapse being in contact with an excitatory synapse

173. Experimentally induced depolarization of excitable membranes has revealed the following regarding membrane excitation:
    1. excitation is decreased at the anode
    2. excitable membranes have no well-defined threshold
    3. depolarization is caused by outward current in the membrane
    4. depolarization is the result of a release of acetylcholine (ACh) in the membrane

174. Spinal shock
    1. is produced by complete spinal transection
    2. is characterized by arc reflexes immediately after transection
    3. duration is often several weeks in humans
    4. is followed by spasticity

175. The major difference between grand mal epilepsy and psychomotor epilepsy is that in a psychomotor seizure there
    1. is no aura
    2. is no postictal state
    3. are myoclonic jerks
    4. is no massive convulsion

176. With a lateral quadrant lesion of the spinal cord, one may observe
    1. paresis, hyperreflexia, hypertonia ipsilaterally
    2. contralateral hypothermia
    3. contralateral anesthesia
    4. ipsilateral Babinski response

Directions Summarized				
A	**B**	**C**	**D**	**E**
1,2,3	1,3	2,4	4	All are
only	only	only	only	correct

177. The thalamic syndrome includes:
    1. emotional changes
    2. various degrees of anesthesia
    3. motor symptoms
    4. decreased threshold to various stimuli

178. Epilepsy
    1. can be genetically linked but is more likely to occur in males
    2. is a disorder of the CNS resulting from paroxysmal cerebral dysrhythmias
    3. can be produced by brain injuries
    4. is most often seen after 30 years of age

179. Sleep and wakefulness are related to which of the following structures?
    1. The intralaminar nuclei of the thalamus
    2. The posterior nucleus of the hypothalamus
    3. The periaqueductal gray
    4. The reticular formation

180. A pituitary adenoma might result in a variety of symptoms (dependent upon the tumor or growth), including:
    1. bitemporal hemianopsia
    2. adrenal hyperfunction (Cushing's disease)
    3. diabetes insipidus
    4. amenorrhea

181. Clinical symptom(s) of cerebellar damage include:
    1. adiadokokinesis
    2. intention tremor
    3. asynergia
    4. ataxia

182. Since atropine is a postganglionic cholinergic-blocking drug, the following physiologic responses are qualitatively common to both stimulation of the sympathetic nervous system and/or systemic administration of atropine:
1. intestinal relaxation
2. bronchiolar dilation
3. increase in heart rate
4. increased secretion of sweat glands

183. Which of the following is common to both excitatory post-synaptic potentials (EPSPs) and inhibitory postsynaptic potentials (IPSPs)?
1. Temporal summation
2. Spatial summation
3. Alteration in Ca^{2+} concentration in the synaptic cleft
4. Postsynaptic membrane depolarization

184. Cortical-evoked potentials
1. is another term for electroencephalograph (EEG) activity
2. are the synchronized recordings after any thalamic stimulation
3. are those responses evoked by direct cortical stimulation
4. are the electrical responses recorded following direct stimulation of either sense organs or afferent fibers that project to the area of cortex under study

185. Regarding cerebellar cortex function
1. it coordinates somatic motor activity and regulates muscle tone
2. after lesions, disturbances are ipsilateral to the lesion
3. no sensory information received by the cerebellum is acted upon at a conscious level by this structure
4. speech is often disrupted after cerebellar damage

Directions Summarized				
A	**B**	**C**	**D**	**E**
1,2,3	1,3	2,4	4	All are
only	only	only	only	correct

186. Regarding sleep and wake mechanisms
    1. the locus coeruleus initiates phasic desynchronization from the deep sleep pattern
    2. lesions of the raphe system induce insomnia
    3. PCO activity is induced by release of a pontine catecholamine pacemaker from inhibition
    4. decreased release of serotonin results in an active hypersomnia

187. In spinal shock
    1. noxious stimuli applied to the skin after spinal transection do not evoke flexion responses
    2. the duration is a function of cerebral dominance
    3. all reflexes served by segments below the spinal transection disappear for a variable amount of time
    4. bladder function is lost

188. Sweat glands are innervated by
    1. cholinergic parasympathetic postganglionic fibers
    2. adrenergic sympathetic postganglionic fibers
    3. cholinergic parasympathetic preganglionic fibers
    4. cholinergic sympathetic postganglionic fibers

189. Clinical evaluation of the peripheral nerves of some alcoholics or vitamin B_1-deficient patients would reveal impairment of
    1. large myelinated fibers
    2. touch, pressure, vibration, and position sense
    3. axolemma but not myelin
    4. small myelinated fibers

190. The following has(have) been contemplated as the cause(s) of the epileptic foci of neurons:
    1. postsynaptic excitation is increased
    2. the neuronal membrane permeability to ions is changed
    3. the amount of neurotransmitter released is deterred
    4. astrocytes form scar tissue

191. Stimulation of the parasympathetic system causes
    1. myosis
    2. accommodation for near vision
    3. increased HCl secretion in the stomach
    4. increased glandular secretions

    ↓ heart activity

    E

192. Inhibition of fear and loss of emotion are prominent signs after lesion of
    1. mammillary bodies
    2. cerebral motor cortex
    3. cerebral frontal lobes
    4. amygdaloid nuclei and limbic system

    D

193. Which of the following statements are true about Purkinje cells?
    1. These cells are tonically active
    2. They give rise to the only axons leaving the cerebellar cortex
    3. They are the largest cells of the cerebellum
    4. They are always excitatory influences on the deep cerebellar nuclei

194. Which of the following is characteristic of chemical synapses?
    1. Synaptic delay
    2. One-way conduction
    3. Susceptibility to drugs
    4. Summate algebraically

Directions Summarized				
A	**B**	**C**	**D**	**E**
1,2,3	1,3	2,4	4	All are
only	only	only	only	correct

195. The action potential of a nerve membrane
    1. has an inward movement of K^+ on its upward part
    2. has a reversal potential of about $+85$ mV
    3. is followed after a delay by an inward movement of K^+
    4. has as its first active change an inward movement of Na^+

196. Paradoxical sleep
    1. is associated with increased muscle tone
    2. consists of abnormal sleep patterns
    3. usually occurs periodically during a night's sleep
    4. is usually associated with dreaming

197. Alpha and beta receptors are
    1. cholinergic receptors
    2. differentiated on the basis of different sensitivities to norepinephrine and strychnine
    3. differentiated by blockade by atropine and curare
    4. adrenergic receptors

198. The "all-or-none" law applies to which of the following events?
    1. Excitatory postsynaptic potential (EPSP)
    2. Inhibitory postsynaptic potential (IPSP)
    3. Presynaptic inhibition
    4. Nerve action potential

199. The outflow from the spinal cord that is the sympathetic nervous system
    1. ceases to function after section of the upper medulla
    2. contains only adrenergic fibers
    3. contains only cholinergic fibers
    4. includes a ganglionic synapse

200. Alpha waves found in the EEG of a normal person
1. have a frequency of 12-24 cps
2. are associated with restful state
3. are most prominent during intense concentration
4. have a frequency of 8-12 cps

201. The functions of the basal ganglia include:
1. the caudate nucleus and putamen initiate gross motor movement
2. the globus pallidus is important in setting background muscle tone
3. the inhibition of muscle tone if they are all stimulated
4. coordinate fine movements of the digits— *motor cortex*

202. Regarding hippocampal functions
1. they have reciprocal EEG activity with the cerebral cortex
2. stimulation and lesions can produce olfactory hallucinations
3. stimulation while under anesthesia can result in arousal and wakefulness which ceases when the stimulation is turned off
4. bilateral lesions in humans have suggested memory deficiencies

203. Extreme obesity results following lesion of the
1. dorsomedial hypothalamic nucleus
2. lateral hypothalamic nucleus
3. supraoptic hypothalamic nucleus
4. ventromedial hypothalamic nucleus

204. Split brain operations
1. are done routinely on epileptic patients
2. reveal that spatial construction tasks are related to the dominant hemisphere
3. reveal that left hemisphere control of the left hand is extremely refined for the distal musculature
4. are characterized by the contralateral hand being dominant in a task learned by both hands but one hemisphere

	Directions Summarized			
A	**B**	**C**	**D**	**E**
1,2,3	1,3	2,4	4	All are
only	only	only	only	correct

205. The pyramidal tract
    1. is composed solely of axons from pyramidal cells
    2. is a crossed pathway
    3. projects solely to the thalamus
    4. originates from several areas of the cortex, including area 4, frontal lobe, and the parietal lobe

206. In the cerebellar cortex, which of the following synapses are excitatory?
    1. Basket cell on Purkinje cell
    2. Climbing fiber on Purkinje cell
    3. Descending fiber on Purkinje cell
    4. Granule cell on Purkinje cell

207. Cerebral blood flow is altered according to which of the following conditions?
    1. Increased physical activity enhances flow
    2. Increased mental activity enhances flow
    3. Changes with local metabolic needs
    4. Increased CO_2 or H^+ concentration causes cerebral vasodilation

208. Frontal lobe lesions result in
    1. euphoria in some cases
    2. reduction in intellectual ability
    3. signs of complacency, self-satisfaction, and often boastfulness
    4. impaired power of judgment of one's situation and narrowing of horizons or goals to the present

209. Neuronal inhibition of afferent fibers
1. is a prominent feature of the dorsal column somatosensory system
2. can be achieved by presynaptic inhibition
3. can be achieved by postsynaptic inhibition
4. is found in dorsal root ganglia

210. Glycine
1. hyperpolarizes motoneurons
2. is blocked by strychnine
3. is present in the terminals of interneuron projections on alpha neurons
4. increases membrane conductance to Cl^- and/or K^+

211. Maximum propagation velocity of the action potential will
1. be observed in myelinated fibers
2. excite skeletal muscle
3. be observed in large-diameter fibers
4. be associated with increased membrane permeability characteristics

212. The hypothalamus is associated with
1. food intake
2. perception
3. water control
4. appropriate integration and control of cardiovascular regulation

213. The limbic system is involved in
1. olfaction
2. feeding behavior
3. rage and fear
4. sexual behavior

		Directions Summarized		
A	**B**	**C**	**D**	**E**
1,2,3	1,3	2,4	4	All are
only	only	only	only	correct

214. Neuronal synapses release ACh that is released from
1. motoneuron terminals in contact with skeletal muscles
2. preganglionic nerve terminals to excite postganglionic neurons of the autonomic nervous system (ANS)
3. preganglionic nerve terminals to excite postganglionic neurons in contact with the kidney juxtaglomerular cells
4. neuronal terminals in the adrenal medulla

215. Fluid ingestion can be increased by
1. increased effective osmotic pressure of the plasma
2. psychologic factors
3. injections of hypertonic saline into the anterior hypothalamus
4. decreased extracellular fluid volume

216. Sleep deprivation is likely to cause
1. striking psychologic effects
2. sluggishness of thought
3. psychotic episodes
4. no affect on body function

217. Dopamine is
1. related to parkinsonism as evidenced by the dopamine content of the caudate nucleus and putamen as being about 50% normal
2. related to prolactin secretion because it is inhibitory and has been used in the treatment of conditions in which there is abnormal milk secretion
3. involved in pathogenesis of schizophrenia because amphetamines stimulate dopamine secretion producing a psychosis that resembles schizophrenia when administered in large doses
4. perhaps the prolactin-inhibitory hormone because it has been found in portal hypophyseal blood

218. Lesions that produce complete inhibition of fear responses and loss of emotion can often be seen in lesions involving the
1. aqueduct of Sylvius
2. olfactory lobes
3. amygdaloid nuclei
4. sensory cortex

219. Recurrent inhibition (inhibitory system) is due to
1. Renshaw cells which receive recurrent collaterals of motoneurons and inhibit other motoneurons in the vicinity
2. a major method of lateral inhibition utilized by neurons throughout the CNS
3. an inhibitory mechanism to sharpen or focus motor output
4. an inhibitory system of the cerebellum

220. Hypothermic cooling for cardiac surgery is utilized because
1. circulation can be stopped for relatively long periods
2. blood pressure is low
3. bleeding is minimal
4. respiration is slowed

221. Which of the following is not a characteristic of spinal cord gray matter organization?
1. Signals ascend to level of the brain stem but no higher
2. Possesses several million neurons per segment of the spinal cord
3. Contains interneurons only
4. Contains interneurons and anterior motor neurons

222. More than half of the fibers descending and ascending the spinal cord
1. provide multisegmental reflex pathways
2. are referred to as propriospinal fibers
3. include pathways for reflex coordination of simultaneous movement of body parts
4. are involved in nociception

Directions Summarized				
A	**B**	**C**	**D**	**E**
1,2,3	1,3	2,4	4	All are
only	only	only	only	correct

223. Which of the following are located in the anterior horn of the spinal cord?
    1. Anterior motor neurons
    2. Interneurons
    3. Gamma motor neurons
    4. Alpha motor neurons

224. Changes to the basal ganglia result in many clinical syndromes. Which of the following is(are) a syndrome(s) associated with basal ganglia damage or disease?
    1. Athetosis
    2. Hemiballismus
    3. Chorea
    4. Dysmetria

225. A typical neuron may be characterized accordingly:
    1. consists basically of two parts: dendrites and the axon
    2. the synaptic afferent fibers are located on the dendritic process which is the cell's input zone
    3. the axon is not electrically excitable and therefore exhibits no action potential
    4. the dendrites are not electrically excitable and therefore exhibit no action potential

226. Which is a feature of the central nervous system of vertebrates?
    1. Telencephalon
    2. Diencephalon
    3. Cerebellum
    4. Brain stem

227. Which of the following may be clinical abnormalities associated with diseases of the cerebellum?
    1. Dysmetria
    2. Ataxia
    3. Dysdiadochokinesia
    4. Dysarthria

228. In the perception of pain
    1. the sensation associated with stimuli that leads to tissue damage is referred to as nociception
    2. the relationship between heat transfer to skin and the pain response is linear
    3. the pain stimuli has inhibitory affects on touch and temperature
    4. the neural basis for pain as suggested by the pattern theory is that specific nociceptive transducers exist

229. The excitatory postsynaptic potential (EPSP) recorded from the cell body of a CNS neuron
    1. is an all-or-none response to a presynaptic impulse
    2. can be temporally summated during repetitive presynaptic stimulation
    3. lasts only for the duration of the presynaptic action potential
    4. can be spatially summated during repetitive firing of several neurons

230. In general, opioids
    1. have inhibitory effects on neurons
    2. have disinhibitory effects on bulbospinal neurons
    3. mimic the effects of endogenous opioids
    4. unlike aspirin, act at the level of transduction

231. Which of the following is (are) implicated in the initiation of a Grand Mal attack?
    1. drugs
    2. fever
    3. loud noises
    4. flashing lights

		Directions Summarized		
A	**B**	**C**	**D**	**E**
1,2,3	1,3	2,4	4	All are
only	only	only	only	correct

232. Which of the following functions are attributable to the level of the spinal cord and/or lower brain?
    1. Walking motions
    2. Reflex control of blood vessels
    3. Equilibrium
    4. Subconscious activities

233. Which of the following are considered to be endogenous opiates of the CNS?
    1. Dynorphin
    2. Beta-endorphin
    3. Beta-lipoprotein
    4. Leu-enkephalin

234. Which of the following is considered part of the analgesia system?
    1. Periaqueductal gray matter
    2. Periventricular nuclei of the hypothalamus
    3. Raphe magnus nucleus
    4. Lateral spinothalamic tract

235. Sensory nerves terminating in the gray matter of the spinal cord
    1. have facilitory effects
    2. enter the cord through the sensory roots
    3. elicit reflex responses
    4. enter the cord through the corticospinal tract

236. Receptive field properties of the striate cortical cells include:
    1. orientation selectivity
    2. binocular receptive fields
    3. direction selectivity
    4. motion parallax

237. Regarding spatial resolution for touch
   1. it is smaller for the hand than for the back
   2. cutaneous receptive fields are larger where more acute discrimination is needed
   3. cutaneous receptive fields are smaller where more acute discrimination is needed
   4. is unaffected by afferent inhibition

238. Deep sleep is or may be
   1. the result of a nearly complete lack of input into the cortex from the reticular activating system
   2. the result of synaptic fatigue
   3. signaled by the appearance of very high-voltage, low-frequency waves on the EEG
   4. associated with a decrease in vegetative functions of the body

---

DIRECTIONS (Questions 239–264): Each of the questions or incomplete statements below is followed by five suggested answers or completions. Select the **one** that is **best** in each case.

---

239. The capacity to display rage
   A. is eliminated when the cerebral cortex is removed
   B. is due to an imbalance of activity in large and small fibers
   C. is not affected by removal of the hypothalamus
   D. does not require any structure above the level of the hypothalamus
   E. is the major function of the ANS

240. The decrease in magnitude of a generator potential during sustained stimulation is called
   A. receptor perturbation
   B. adaptation
   C. refractoriness
   D. accommodation
   E. chronaxie

241. Normal blood flow to the brain is
    A. greatly modified by vasomotor control
    B. increased by high O_2 level
    C. about 150 mL/min
    D. about 750 mL/min
    E. greatly increased during exercise

242. Synaptic innervation of a number of cells by one fiber is an example of
    A. convergence
    B. chronaxie
    C. rheobase
    D. divergence
    E. reverberation

243. Activation of regional areas of the cortex
    A. is accomplished by the reticular activating system
    B. is accomplished by the diffuse thalamocortical system
    C. is induced only by painful stimuli
    D. may be associated with the direction of our attention to one area of our environment
    E. is determined solely by blood flow

244. Fever due to infection is the result of
    A. increased heat production followed by decreased heat loss
    B. decreased heat loss followed by increased heat production
    C. an action of bacterial pyrogens independent of the body's heat control mechanism
    D. a functional hyperthyroidism caused by irritation of the thyroid and release of colloid
    E. increased respiratory work

245. The vomiting center is located in the
    A. cerebral cortex
    B. thalamus
    C. hypothalamus
    D. medulla oblongata
    E. cervical spinal cord

246. The inhibitory synaptic potential recorded from the cell body of a CNS neuron
    A. cannot be summed either temporally or spatially
    B. involves a selective increase in the permeability to K^+ and Cl^-
    C. cannot be opposed by stimulation of excitatory pre-synaptic neurons
    D. is not a normal occurrence during spinal reflex activity
    E. involves a selective increase in Ca^{2+} permeability

247. During the excitation of a nerve cell the peak of potassium efflux occurs
    A. after the spike and before the peak sodium influx
    B. before the spike and after the peak sodium influx
    C. before both spike and peak sodium influx
    D. after both spike and peak sodium influx
    E. coincident with the peak of sodium influx

248. The body temperature range that can be tolerated best during hypothermic cooling is (in °F)
    A. under 69°
    B. 70°–74°
    C. 75°–79°
    D. 80°–84°
    E. 85°–90°

249. The primary motor cortex
    A. receives no sensory input
    B. is active in the adjustment of motor activity to current sensory input
    C. is not necessary for fine motor movement
    D. gives rise to the extrapyramidal tract
    E. is localized only in the frontal lobe

250. Norepinephrine is probably the
    A. inhibitory transmitter at the alpha motoneuron
    B. excitatory transmitter acting on Renshaw cells
    C. transmitter at muscarinic synapses
    D. transmitter released from most postganglionic sympathetic fibers
    E. transmitter released from Purkinje cell synapses

251. The sympathetic division of the autonomic nervous system is characterized by
    A. presynaptic inhibition
    B. thoracolumbar outflow from the spinal cord
    C. short postganglionic fibers
    D. adrenergic preganglionic fibers
    E. the vagus nerve, which is its major component

252. Premotor cortex refers to
    A. some areas anterior to the primary motor cortex that can cause complex coordinate movements, such as speech, eye, and head movements
    B. an area of the motor cortex that is vital for the initiation of voluntary motor movements
    C. an area found in the temporal cortex that when stimulated allows the most primitive movements to take place
    D. a term often used to refer to the cerebellar cortex
    E. an area of the cortex in the vicinity of the insula

253. The motor cortex
    A. could be considered as six separate motor areas that when stimulated are especially likely to cause contraction of specific muscles
    B. can be considered to be composed of only small neurons
    C. is composed of primary motor, supplementary motor, somatic sensorimotor I, and somatic sensorimotor II areas
    D. can be found completely in the frontal lobe
    E. has no apparent organization of motor functions

254. The human rectal temperature at which permanent cell damage might result if it is prolonged is (in °F)
    A. 99°
    B. 101°
    C. 103°
    D. no effect is observed even at 110°
    E. 106°

255. The release of transmitter from nerve terminals into the synaptic cleft
    A. decreases dramatically when the nerve terminals are depolarized
    B. is associated with the influx of Ca^{2+} into the presynaptic membrane during an action potential in the nerve endings
    C. in a normal individual will always raise the postjunctional membrane above threshold if an action potential invades the nerve terminal
    D. is inhibited if the nerve terminals are hyperpolarized
    E. is an all-or-none phenomenon

256. Activation of various portions of the reticular formation
    A. can increase reflex activity
    B. can cause a complex motor movement such as speech
    C. can decrease reflex activity
    D. cannot affect the reflex activity
    E. acts to modulate reflex activity in conjunction with other brain structures

257. The condition known as REM (rapid eye movement) sleep is
    A. that point at which the individual becomes aware and alert
    B. characterized by slow high-voltage regular EEG activity
    C. referred to as paradoxical sleep
    D. related to EEG patterns seen in comatose patients
    E. characterized by total lack of all muscular activity

258. The cerebellum
    A. has a totally inhibitory output from its cortex
    B. has an excitatory output from its deep nuclear layers
    C. receives cortical input from mossy and climbing fibers
    D. has the same arrangement of cells as in most of the cerebellar cortex
    E. has a conscious interpretation of motor activity

259. A pair of electrodes placed on the surface of an uninjured nerve will record (following nerve stimulation)
    A. a resting potential followed by a monophasic compound action potential
    B. a zero potential followed by a diphasic compound action potential
    C. a zero potential followed by a monophasic compound action potential
    D. a resting potential followed by a diphasic compound action potential
    E. no potential change at all

260. At the postsynaptic membrane the EPSP is
    A. produced by ACh, giving rise to an increased permeability first to Na^+ and then after a delay to K^+
    B. produced by ACh, causing an increased permeability to Na^+ and K^+ simultaneously
    C. caused by a permeability increase to all ions except Na^+
    D. brought about by the splitting of ACh
    E. produced by ACh, giving rise initially to an increased K^+ permeability and then after a delay to Na^+

261. Metabolism in the nerve fiber
    A. supplies ATP to the sodium pump
    B. is blocked by cyanide or dinitrophenol with a directly correlated fall in resting and action potentials
    C. does not supply energy for fast transport of material
    D. controls the ACh level and, thus, the action potential
    E. is determined by its position in a nerve bundle

262. The cerebellum
    A. is associated with very rapid motor activity
    B. may be a timing device for measuring duration of rapid motor activity
    C. receives input from most of the cerebral cortex
    D. is only activated by painful stimuli
    E. is only associated with unlearned motor movements

263. Gamma-aminobutyric acid (GABA)
    A. hyperpolarizes motoneurons and is inhibitory in nature
    B. is blocked by strychnine
    C. is blocked by glycine
    D. probably is responsible for the inhibitory postsynaptic potential (IPSP) in alpha motoneurons
    E. is excitatory in nature

264. Alpha receptors differ from beta receptors in that
    A. alpha receptors are generally inhibitory, while the beta receptors are generally excitatory
    B. norepinephrine interacts with alpha receptors but not with beta receptors
    C. alpha stimulation is generally followed by prolonged desensitization of the effector structures
    D. alpha receptors are found only in glands
    E. they allow an affinity of a hormone to a given organ

Explanatory Answers

170. E. The medullary center is under the influence of chemical changes in the blood, blood pressure alterations, afferent impulses from the viscera, and higher control centers. White blood cell concentration in the blood has no direct effect upon the centers. (REF. 2, p. 978 ff)

171. E. When excited, the eating centers that are associated with hunger are the perifornical and lateral nucleus of the hypothalamus. Damage to these areas causes the animal to lose desire for food. A center that looses the desire for food is called the satiety center. These hypothalamic nuclei may be served by glucose receptors. (REF. 2, p. 1546 ff)

172. E. Presynaptic inhibition is caused by the presence of inhibitory knobs lying directly on the terminal fibrils and excitatory knobs themselves. They secrete a transmitter substance that partially depolarizes the terminal fibrils and the excitatory synaptic knobs. Consequently, the voltage of the action potential that occurs at the membrane of the excitatory knob is depressed, and this greatly reduces the amount of excitatory transmitter released at the knob. Therefore, the degree of excitation of the neuron is greatly depressed. (REF. 3, p. 58 ff)

173. B. At the cathode, the potential outside the membrane is negative with respect to that on the inside. Current flows outward through the anode. Cathodal current decreases, while an anodal current actually increases resistance to excitation. The all-or-none principle of excitation is based on the finding that electrical stimulation has a strength limit above which a full action potential results with propogation. (REF. 2, p. 12 ff)

174. E. With time, the depression of segmental reflexes disappears. Usually, the first reflexes to return after a period of spinal shock are the flexion reflexes mediated by polysynaptic pathways. (REF. 2, p. 737 ff)

175. D. One type of focal epilepsy is the so-called psychomotor seizure, which may cause (1) a short period of amnesia; (2) an attack of abnormal rage; (3) sudden anxiety, discomfort, or fear; (4) a moment of incoherent speech or mumbling of some trite phrase; or (5) a motor act to attack someone, to rub the face with the hand, or so forth. (REF. 2, p. 715 ff)

176. E. Examination of a patient who has sustained unilateral anterolateral cordotomy reveals several sensory changes. First, contralateral loss of pain and temperature sensation. Motor components would reflect a typical upper motoneuron lesion set of symptoms. (REF. 2, p. 718 ff)

177. E. The thalamic syndrome involves emotional changes, various degrees of anesthesia, and motor symptoms such as ataxia. There is also a permanent loss or severe impairment of light touch and position as well as a delay in the recognition of superficial stimuli. No sexual functions, however, are altered in thalamic lesions. (REF. 2, p. 332)

178. A. The usual course is for petit mal attacks to appear in late childhood and then to disappear entirely by the age of 30. Petit mal will often become grand mal, since both essentially originate in the same locus. (REF. 2, p. 715 ff)

179. E. Sleep or sleeplike behavior can be produced by stimulation of the thalamic intralaminar nuclei, the preoptic and supraoptic areas, and the caudate. The existence of a wakefulness center is questionable. The ventral posterolateral nucleus of the thalamus has no relationship to sleep. (REF. 2, p. 1239 ff)

180. E. Some of the hypothalamic lesion syndromes include the hyperthermic syndrome, the diabetes insipidus and emaciation syndrome, the adiposogenital dystrophy syndrome, the syndrome characterized by somnolence and disorder of temperature regulation, and finally diencephalic or autonomic epilepsy. In no case will psychosis be found in hypothalamic lesions. (REF. 3, p. 895 ff)

181. E. The common denominator of most cerebellar signs is inappropriate rate, range, force, and direction of movement. Lack of reflexes (areflexia) is not a cerebellar symptom. (REF. 2, p. 641 ff)

182. A. Stimulation of the nervous system results in increased, copious sweating, whereas with atropine there is a decreased secretion of sweat glands. (REF. 3, p. 280 ff)

183. A. Release of neurotransmitter into the synaptic cleft may result in either of two responses, inhibition or excitation of the postsynaptic membrane. An increase in intraneuronal voltage from the normal resting potential (i.e., resting membrane potential becomes more negative) caused by a neurotransmitter, indicates a hyperpolarized state. This is called inhibitory postsynaptic potential (IPSP). Because of the structural and integrative properties of neurons, synaptic inputs (both EPSPs and IPSPs) can occur by either spatial or temporal summation. In either case there is a change in the concentration of calcium ions, which mediates the release of neurotransmitter in the synaptic cleft. Postsynaptic membrane depolarization is indicative of EPSPs. (REF. 3, p. 55 ff)

184. D. The electrical events that occur in the cortex after stimulation of a sense organ can be monitored with an exploring electrode connected to the reference electrode. The first positive–negative wave sequence is the primary evoked potential; the second is the diffuse secondary response. (REF. 2, p. 718 ff)

185. E. Care should be taken to avoid damaging the nuclei when surgical removal of parts of the cerebellum is necessary. Compensation for the effects of cortical lesions occurs, but compensation for the effects of lesions of the cerebellar nuclei does not occur. (REF. 2, p. 641 ff)

186. A. True active hypersomnia with accompanying slow-wave sleep and paradoxical sleep is due to increased release of serotonin. (REF. 2, p. 1251 ff)

187. E. Usually the first reflexes to return after a period of spinal shock are the flexion reflexes mediated by polysynaptic pathways. (REF. 3, pp. 199, 248 ff)

188. D. Sweat glands are innervated by sympathetic cholinergic postganglionic fibers. (REF. 2, p. 751 ff)

189. D. The peripheral neuropathy associated with alcoholism and deficiency of vitamin B_1 selects small fibers and produces a modality dissociation. (REF. 3, p. 735)

190. E. Recent studies in simple vertebrates have shown that the neuroglia are almost exclusively permeable to potassium ions. The potential gradient is established between depolarized glia, where extracellular potassium is high, and adjacent glia cells. The resulting current flow through bridges of low resistance will tend to remove potassium from clefts where the concentration of this ion is high. This method of buffering the extracellular potassium concentration is one possible role of the glia scar formation. These sites are often areas of epileptic activity. Thus, although there is an alteration in ionic flow, astrocytes are not producing the neuronal anoxia. (REF. 1, p. 678 ff)

191. E. Parasympathetic stimulation decreases the overall activity of the heart. (REF. 3, p. 411)

192. D. After destruction of the amygdaloid nucleus and lumbar system, the normal fear reaction is often absent. (REF. 3, p. 295)

193. A. Purkinje cells are always inhibitory influences on the tonically active neurons of the cerebellar nuclei. (REF. 2, p. 632 ff)

194. E. Because a synapse represents a physical "gap" between presynaptic and postsynaptic membranes, there is a time barrier which is a function of that distance. The time required for a neurotransmitter to traverse this distance is referred to as synaptic delay. Unlike axonal transmission of action potentials, which can travel in both directions along its length, synaptic transmission is in one direction, due to the presynaptic and postsynaptic arrangement of secretory vesicle and receptors, respectively. Furthermore, axonal potentials exhibit the all-or-none principle while synapses summate algebraically, adding excitatory while subtracting inhibitory potentials. (REF. 2, p. 236 ff)

195. D. When the sodium channels open and these ions pour into the side of the membrane, the positive charges of the sodium ions neutralize the normal electronegativity inside the fiber and also create an excess of positive charges. Hence, the membrane potential inside the fibers is positive and is called the reversal potential at about the time K^+ permeability increases and repolarization is initiated. (REF. 3, p. 33)

196. D. Paradoxical sleep is usually associated with active dreaming. (REF. 3, p. 267)

197. D. Drugs that mimic the action of norepinephrine on sympathetic effector organs have shown that there are at least two different types of adrenergic receptors. (REF. 2, p. 737 ff)

198. D. Under the same resting conditions a nerve action potential will always have the same amplitude. (REF. 2, p. 57 ff)

199. D. A sympathetic path always contains a preganglionic and postganglionic neuron. (REF. 3, p. 280)

200. C. Alpha waves are seen most often in a resting but awake person. They occur 8 to 12 times per second. (REF. 3, p. 266)

201. A. Fine control of the digits is dependent on the function of the primary motor cortex. (REF. 3, pp. 241, 253 ff)

202. E. Many investigators who have studied behavioral effects in the monkey consequent to bilateral destruction of the amygdaloid nuclei, uncus, or hippocampi have found that the animals were tame, fearless, and asocial for 4 to 5 months. (REF. 2, p. 704 ff)

203. D. Destruction or anesthetization of the ventromedial nuclei leads to acceleration of self-stimulation and feeding. (REF. 2, p. 1545 ff)

204. D. Split brain operations on humans have revealed that in a task learned by a single hemisphere, but where both hands are part of the learned task, the contralateral hand will be dominant. (REF. 2, p. 682 ff)

205. C. The pyramidal tract is a crossed pathway with a diffuse origin. (REF. 3, pp. 215, 250 ff)

206. C. Climbing fibers and granule cells are excitatory to the Purkinje cell in the cerebellar cortex. (REF. 2, p. 568 ff)

207. E. Adjustment of cerebral blood flow in accordance with local requirements of cerebral metabolism is well documented. Increases in carbon dioxide or hydrogen ion concentration will dilate cerebral arterioles and thereby increase flow. (REF. 2, p. 952 ff)

208. E. After frontal lobotomy many patients exhibit tactlessness, extroversion, euphoria, and noticeable liability of emotions with a tendency to outbursts. Some observers report a definite lowered intellectual ability, and nearly all agree that the capacity for abstract thinking is lessened. There is a noticeable distractibility in the majority of patients. Olfactory hallucinations have no relation to the frontal lobe. (REF. 2, p. 679 ff)

209. E. Afferent inhibition is a prominent feature of the dorsal column somatosensory system. It can be achieved by presynaptic and postsynaptic inhibition. A volley of impulses in dorsal root afferents leaves in its wake an enduring depolarization of the intraspinal segments of both the active and the adjacent fibers. (REF. 1, p. 556 ff)

210. E. Certain characteristics of physiologic inhibitory transmission, including true hyperpolarization and increased membrane conductance to Cl^- or K^+, that are antagonized by strychnine and related compounds have been produced by glycine. Glycine is an inhibitory transmitter. (REF. 2, p. 246 ff)

211. A. The velocity of conduction in nerve fibers varies from as little as 0.5 m/s in very small unmyelinated fibers up to as high as 130 m/s in very large myelinated fibers. The velocity increases proximately with the fiber diameter in myelinated nerve fibers and proximately with the square root of fiber diameter in unmyelinated fibers. The excitation of nodes permits the saltatory conduction and muscle excitation. (REF. 2, p. 58 ff)

212. E. Unquestionably the hypothalamus is of importance to the perception of thirst and the control of water intake. It is also involved in the regulation of body water loss, partially via the kidney. (REF. 2, p. 697 ff)

213. E. Stimulation and ablation experiments indicate that in addition to its role in olfaction, the limbic system is concerned with feeding behavior. Along with the hypothalamus, it is also concerned with sexual behavior and the emotions of rage, fear, and motivation. (REF. 1, p. 697 ff)

214. E. Hypothalamic releasing factors are related to portal vessels for control of anterior pituitary hormones. All sites listed are under the influence of their transmitter. (REF. 1, p. 689 ff)

215. A. Drinking can be increased by increased effective osmotic pressure of the plasma, by decreases in ECF volume, and by psychologic and other factors. Injections of hypertonic saline into the anterior hypothalamus causes drinking in conscious animals. (REF. 1, p. 432 ff)

216. A. Prolonged wakefulness is often associated with progressive malfunction of the mind and behavioral activities of the nervous system. A person can become irritable or even psychotic following forced wakefulness for prolonged periods of time. (REF. 1, p. 1259 ff)

217. E. The physiologic role of dopamine is related to parkinsonism because dopamine content of the caudate nucleus and putamen is about 50% normal. It is related to prolactin secretion because it is inhibitory and has been used in the treatment of conditions in which there is abnormal milk secretion. Pathogenesis of schizophrenia has been postulated because amphetamines, which stimulate dopamine secretion, produce a psychosis that resembles schizophrenia when administered in large doses. It may actually be the prolactin-inhibitory hormone because it has been found in portal hypophyseal blood. It is not related to the opiate receptors or pain. (REF. 1, p. 553 ff)

218. B. The fear reaction and its autonomic and endocrine manifestations are absent in situations in which they would normally be evoked when the amygdala is destroyed as well as the aqueduct of Sylvius. (REF. 1, pp. 680, 684)

219. A. Small interneurons, called Renshaw cells, lie in close association with motoneurons. Motoneuron axon collaterals pass to Renshaw cells, which in turn transmit inhibitory signals to nearby motoneurons. Both sensory and motor systems utilize lateral inhibition to sharpen and focus their signals. (REF. 1, p. 606 ff)

220. E. When the skin or blood is cooled enough to lower the body temperature in humans metabolic and physiologic processes slow down. Respiration and heart rates are very slow, blood pressure is low, bleeding is minimal, and consciousness is lost. These conditions are excellent for surgery. (REF. 1, p. 860)

221. C. In each segment of the spinal cord there are several million neurons in its gray matter. Each segment of the anterior horn of gray matter contains several thousand neurons 50% to 100% larger than most others. These are the anterior motor neurons. Included are interneurons. The neurons of gray matter may travel to higher levels in the cord itself, to the brain stem, or even to the level of the cerebral cortex. (REF. 1, pp. 606–607)

222. A. The multiple segmental interconnecting fibers of the spinal cord are referred to as the propriospinal fibers and comprise more than half of the fibers of the spinal cord. Both ascending and descending, they include reflexes for coordinated control of the simultaneous movement of fore and hind limbs. (REF. 1, p. 607)

223. E. All of the neurons listed are located in all areas of the gray matter, including the anterior horn. (REF. 1, p. 607)

224. A. Chorea is characterized by random uncontrolled "flicking" movements. Hemiballismus is characterized by violent, uncontrolled movements of larger areas of the body. They are usually successive, occurring once every few seconds or only once in many minutes. Athetosis is slow "wormlike" writhing movements of the neck, face, and hands. Dysmetria is a condition associated with diseases of the cerebellum. (REF. 1, p. 629)

225. C. The neuron has three distinct regions: dendritic processes, soma, and axon. The dendrites represent the input zone and the axon represents the cells sole means of output. The soma and dendrites are not electrically excitable in that they do not display voltage-dependent conductance and therefore do not display an action potential. On the other hand the axon displays all these features. (REF. 3, p. 71)

226. E. Another component is the spinal cord. (REF. 3, p. 73)

227. E. All listed. (REF. 1, p. 647)

228. A. There are theories, including the pattern theory and gate control theories, for the pain response. The most widely accepted is the Specific theory, which suggests that there are specific nociceptive transducers that are actually responsible for the pain response. However, the perception of pain may include aspects of all three theories. It is also known that pain can have inhibitory effects on touch and temperature perception. (REF. 3, pp. 144–145)

229. C. When an afferent volley to the cerebral cortex is either purely excitatory or purely inhibitory for a cell under stimulation, that cell responds with EPSPs that are depolarizing or hyperpolarizing. These can summate spatially and/or temporally but it is not an all or none response.

230. A. Aspirin, a true analgesic, works at the level of transduction. Opioids like morphine act more centrally and mimic the endogenous opioids. They have, in general, inhibitory effects on neurons. However, the bulbospinal neurons block nociception when activated. Therefore, it is believed that opioids activate the bulbospinal neurons by disinhibition. (REF. 3, pp. 151–152)

231. E. All the factors listed can initiate a Grand Mal attack in those individuals so predisposed. However, in individuals not genetically predisposed, traumatic lesions in almost any part of the brain can cause excess excitability in local brain areas. These may transmit signal into the reticular activating system eliciting a Grand Mal attack. (REF. 1, p. 674)

232. E. Functions associated with the level of the spinal cord include withdrawal reflexes and support against gravity but equilibrium and subconscious activities are functions of the lower brain level: medulla, cerebellum, and basal ganglia. (REF. 1, p. 548)

233. E. All listed. (REF. 1, p. 597)

234. A. The lateral spinothalamic tract is part of the anterolateral system for transmission of sensory signals that do not require discrimination of intensity or discrete localization. (REF. 1, p. 589)

235. A. The sensory signal enters the spinal cord through the sensory roots. Their effects include local segmental and excitatory responses, facilitory effects, and reflexes. (REF. 1, p. 606)

236. A. Motion parallax is a phenomenon associated with monocular cues of depth, specifically, the retinal image of objects moving at different rates and at different distances. (REF. 3, pp. 122–123)

237. B. Spatial resolution is defined as the minimal separation for two stimuli to be distinguished as one. Where tactile acuity is more important, receptive fields are smaller, such as for the hand, and where tactile acuity is not so important, receptive fields are larger, as for the body trunk. Unlike primary cutaneous afferent fibers, neurons of the dorsal column, somatosensory cortex, and ventral posterolateral nuclei possess lateral inhibition; stimuli in one area inhibits excitation in surrounding zonal regions. (REF. 3, pp. 139–140)

238. E. Deep sleep is characterized by high-voltage delta waves occurring at a rate of 1 to 2 per second. It is dreamless, associated with a decrease in both peripheral vascular tone and most of the other vegetative functions of the body as well. It may be due to synaptic decay and is probably due to a nearly complete lack of input by which a person can direct his attention to specific aspects of his mental environment. (REF. 1, p. 671 ff)

239. D. Studies of animals (that have survived the removal of the cerebral cortex) have shown conclusively that the capacity to display anger or rage reactions depends on subcortical mechanisms.

In studies of both acute and chronic animals, it has been shown that the hypothalamus is necessary for the vigorous expression of rage reactions. (REF. 1, p. 681 ff)

240. B. A special characteristic of all sensory receptors is that they adapt either partially or completely to their stimuli after a period of time. When a continuous sensory stimulus is applied the receptors respond at a very high impulse rate at first, then progressively less rapidly until finally many of them no longer respond at all. (REF. 1, p. 575 ff)

241. D. The normal blood flow through brain tissue averages 50–55 mL/100 g of brain per minute. For the entire brain of the average adult, this is approximately 750 mL/min. (REF. 1, p. 338 ff)

242. D. Dorsal root fibers, upon entering the central nervous system, send branches to many different cells. This divergence makes a single sensory input available to many structures. (REF. 1, p. 564 ff)

243. D. Since the cerebral cortex is one of the most important areas of the brain for conscious awareness of our surroundings, one can surmise that the ability of specific thalamic areas to excite specific cortical regions might be one of the mechanisms by which a person can direct his attention to specific aspects of his mental environment. (REF. 1, p. 634 ff)

244. B. The initial reaction during fever is a decreased heat loss because the body does feel cold. Secondarily, there is an increased heat production. (REF. 1, p. 858 ff)

245. D. The so-called vomiting center is found in the dorsal part of the lateral reticular formation of the medulla oblongata. (REF. 1, p. 803 ff)

246. B. The inhibitory synaptic potential involves a selective increase of K^+ or Cl^- but not to Na^+. The synaptic membrane is then less readily depolarized by the action of excitatory presynaptic terminals. (REF. 1, p. 556 ff)

247. D. During the rising phase of the action potential the major part of the conductance increase is due to sodium. The potassium conductance does not increase appreciably until near the peak of the spike. Thereafter, the sodium conductance decreases rapidly, and a proportionately greater fraction of the total conductance is potassium. (REF. 2, p. 218 ff)

248. E. Humans tolerate body temperatures of 85°–90°F without permanent ill effects. (REF. 1, p. 860 ff)

249. B. The somatic sensory area's relationship to the primary motor cortex displays a close functional interdependence of the two areas. It is primarily in the sensory and sensory association areas that one experiences effects of motor movements and records "memories" of the different patterns of motor movements. These are called sensory engrams of the motor movements. (REF. 1, p. 632)

250. D. The majority of the sympathetic postganglionic endings secrete norepinephrine. (REF. 2, p. 741 ff)

251. B. There is a preganglionic fiber outflow from each segment of the spinal cord from the first thoracic to the third lumbar level. (REF. 1, p. 237 ff)

252. A. Some areas anterior to the primary motor cortex that can cause complex coordinate movements, such as speech, eye movements, and head movements, are referred to as the premotor cortex. (REF. 1, p. 632 ff)

253. C. The motor cortex can be considered as four separate motor areas: the primary motor, supplementary motor, somatic sensorimotor I, and somatic sensorimotor II. (REF. 1, p. 632 ff)

254. E. When the rectal temperature is over 41° C (106° F) for prolonged periods some permanent brain damage results. (REF. 1, p. 896 ff)

255. B. The number of vesicles released with each action potential is greatly reduced when the quantity of calcium ions in the extracellular fluid is diminished. It has been suggested that the spread of the action potential over the membrane of the knob causes small amounts of calcium ions to leak into the knob. The calcium ions then supposedly attract the transmitter vesicles to the membrane and simultaneously cause one or more of them to rupture, thus allowing spillage of their contents into the synaptic cleft. (REF. 2, p. 231 ff)

256. E. By far the majority of the reticular formation is excitatory. Diffuse stimulation of facilitory areas causes a general increase in muscle tone throughout the body or in localized areas. In the normal animal, inhibitory signals are continually available from the basal ganglia, cerebellum, and the cerebral cortex to keep the facilitory system from becoming overactive. (REF. 1, pp. 596, 619)

257. C. EEG patterns become rapid, of low voltage, and irregular, and these resemble the EEGs seen in alert humans. The threshold for arousal by sensory stimuli is elevated. This condition is called paradoxical sleep. (REF. 2, p. 803 ff)

258. A. The inhibitory influences of the deep cerebellar nuclei arise entirely from the Purkinje cells in the cortex of the cerebellum. All of the efferent tracts from the cerebellum arise in the deep nuclei; none arise from the cerebellar cortex. (REF. 1, p. 638 ff)

259. B. A pair of electrodes placed on the surface of an uninjured nerve will record a zero potential followed by a diphasic compound action potential. (REF. 1, p. 117)

260. B. The EPSP is a local depolarization of the postjunctional membrane caused by a transient increase in the ionic permeability of this membrane. The action of ACh on the postjunctional membrane is to increase its permeability to all species of ions. (REF. 2, p. 238 ff)

261. A. Since the pump requires energy for operation, this process of "recharging" the nerve fiber is an active metabolic one, utilizing energy from the ATP energy "currency" system of the cell. (REF. 2, p. 60 ff)

262. C. The cerebellum receives an extensive afferent pathway by the corticocerebellar pathway, which originates mainly in the motor cortex (but to a lesser extent in the sensory cortex as well). It then passes by way of the pontile nuclei and pontocerebellar tracts directly to the cortex of the cerebellum. (REF. 2, p. 632 ff)

263. A. GABA appears to have a true hyperpolarization at the lateral vestibular nucleus and cerebral and cerebellar cortices, except that its action is resistant to strychnine. In the crustacean peripheral inhibitory motor fibers, GABA is already convincingly identified as the inhibitory transmitter. (REF. 1, p. 627 ff)

264. E. Certain alpha functions are excitatory, while others are inhibitory. Therefore, alpha and beta receptors are not associated with excitation or inhibition, but simply with the affinity of the hormones for the receptors in a given organ. (REF. 1, pp. 298, 373 ff)

4 Respiratory Physiology

DIRECTIONS (Questions 265-307): For each of the questions or incomplete statements below, **one** or **more** of the answers or completions given is correct. Select
A if only *1, 2, and 3* are correct
B if only *1 and 3* are correct
C if only *2 and 4* are correct
D if only *4* is correct
E if all are correct

265. Which of the following is true regarding respiration?
1. Influenced by centers located in the medulla
2. Influenced by centers located in the cerebral cortex
3. At the end of inspiration the pressure in the alveolar space within the lungs is subatmospheric
4. At the end of expiration the pressure in the alveolar space within the lungs is subatmospheric

266. CO_2 is carried by the blood
1. in physical solution with the plasma
2. in combination with hemoglobin
3. in the form of bicarbonate
4. bound to non-heme protein

267. Which is NOT a feature influencing O_2 delivery and CO_2 removal from the body?
    1. Changes in intrathoracic pressure during the breathing cycle
    2. Interaction among sensory receptors located in the lungs
    3. Interaction among sensor receptors located in the arteries
    4. The churning action of gas within the lung created by contractions of the heart

268. During contraction of the diaphragm
    1. the intraabdominal pressure becomes positive
    2. the diaphragm flattens
    3. the vertical dimensions of the thoracic cage increase
    4. the lateral dimensions of the thoracic cage decrease
    in

269. The diaphragm
    1. accounts for half the inspired air during inspiration
    2. accounts for less than one third of the inspired air during anesthesia   phrenic
    3. is supplied by the vagus nerve
    ④ is over two-thirds made up of slow-twitch fibers which resist fatigue

270. To determine the cost of breathing
    1. oxygen consumption is measured at rest and at increased ventilation
    2. oxygen consumption is measured before and after exercise
    3. normally less than 5% total O_2 is consumed by the body
    4. oxygen consumption is decreased as the elasticity of the chest bellows increases

271. Functional residual capacity is
    1. the volume of air inhaled during inspiration
    2. the volume of air exhaled during exhalation
    3. the maximal amount of air that can be exhaled after quiet expiration
    4. the amount of air left in the lungs at the end of normal, resting expiration

Directions Summarized				
A	B	C	D	E
1,2,3 only	1,3 only	2,4 only	4 only	All are correct

272. The volume of air contained in the lungs
   1. after maximal expiration is the residual volume
   2. after maximal inspiration is termed expiratory reserve volume (ERV)
   3. after maximal inspiration from functional residual capacity is inspiratory capacity (IC)
   4. none of the above

273. Which of the following is true regarding the elastic properties of the lung?
   1. Collagen and elastin are responsible for the resilience of the lung
   2. In diseases that decrease distensibility the elastic recoil exceeds that of the normal lung
   3. Emphysema increases the distensibility of the lung
   4. Emphysema decreases the distensibility of the lung

274. The elastic recoil of the chest wall
   1. is directed inward at end inspiration
   2. is directed outward at functional residual capacity
   3. is opposed by the recoil of the lungs
   4. is approximately 70% of lung capacity at equilibrium

275. Fluid flow through the circulatory system is in many respects similar to fluid flow through a tube in that
   1. it is characterized by laminar flow
   2. stream lines are faster near the center of flow
   3. flow rate is inversely proportional to length of the tube
   4. it is directly proportional to the square of the diameter of the tube

276. Which of the following is (are) true regarding air and water flow through a tube?
   1. Flow rate is proportional to the square root of the driving pressure
   2. Both show a parabolic profile during laminar flow
   3. Molecules are free to collide more randomly and therefore form eddies much more readily
   4. Fluid flow rate is independent of gas density

277. Which of the following is true regarding distribution of airway resistance in respiration?
   1. Greater during nose breathing than mouth breathing
   2. Greater in the subsegmental regions than in upper trachea or bronchial tree
   3. Airway resistance decreases with each successive branching
   4. Total cross-sectional area increases with each successive branching

278. At the beginning of forced expiration
   1. lung elastic recoil increases
   2. airway resistance decreases
   3. intrathoracic airways widen
   4. transmural airway pressure decreases

279. During a maximal inspiratory event
   1. the pleural pressure is subatmospheric
   2. force generation of the muscles of inspiration increases
   3. flow rate remains high even over a wide range of lung volumes
   4. the transmural pressure remains low

280. Venous CO_2 is in the form of
   1. carbamino compounds
   2. bicarbonate
   3. dissolved CO_2
   4. carbonic acid

		Directions Summarized		
A	**B**	**C**	**D**	**E**
1,2,3	1,3	2,4	4	All are
only	only	only	only	correct

281. The O_2 dissociation curve is shifted to the right by
    1. increased temperature of blood flow
    2. increased pCO_2
    3. decreased pH
    4. increased N_2 DPG

282. Alveolar space pressure is
    1. subatmospheric at the end of inspiration
    2. less than atmospheric pressure during expiration as a consequence of lung compliance
    3. independent of airflow direction and airway dimensions
    4. dependent upon airway dimensions and direction of airflow

283. Severe hypoventilation will result from
    1. systemic arterial blood hypocapnia
    2. direct stimulation of medullary chemoreceptors by hypoxia
    3. an abnormally low alveolar pCO_2
    4. cyanosis and alkalosis of systemic arterial blood

284. Hyperpnea resulting from moderate exercise may be due to
    1. increased body temperature
    2. joint movement
    3. serum pH decrease
    4. increase in serum pCO_2

285. The diffusion of O_2 across the alveolar membrane is much less than that of CO_2 because
    1. CO_2 is actively transferred
    2. CO_2 is more soluble in H_2O (than is O_2), which enables it to pass the membrane easier
    3. the alveolar area available for O_2 diffusion is larger
    4. the CO_2 molecular weight is greater than O_2

286. Loss of or decreased vagal activity results in a(n)
    1. decreased rate of respiration
    2. increased depth of respiration
    3. loss of some sensory input to the respiratory centers
    4. more irregular breathing pattern

287. A patient presents with severe prolonged vomiting from deep inside the gastrointestinal (GI) tract. The most likely consequence(s) is (are)
    1. decrease in plasma pH
    2. metabolic acidosis
    3. increase in plasma bicarbonate
    4. increase in rate and depth of respiration

288. Hemoglobin is particularly well suited to carry oxygen in the blood. The advantages seen include:
    1. an easily reversible binding of O_2
    2. a greater affinity for O_2
    3. the ability to give up most of its oxygen at pO_2 between 10 and 80 mm Hg
    4. the ability to give up more O_2 to actively metabolizing tissue

289. O_2 release from hemoglobin is caused and enhanced by
    1. low pH in the tissues
    2. high temperature in the tissues
    3. high pCO_2 in the tissues
    4. low pO_2 in the tissues

290. Carbon monoxide
    1. loosely combines with CO_2 in the plasma
    2. interferes with O_2 transport
    3. interferes with CO_2 transport
    4. has a greater affinity to combine with hemoglobin than does O_2

291. Hering-Breuer reflexes result in
    1. inhibition of inspiration when the lungs are inflated
    2. excitation of inspiration when the lungs are inflated
    3. excitation of inspiration when the lungs are deflated
    4. inhibition of inspiration when the lungs are deflated

		Directions Summarized		
A	**B**	**C**	**D**	**E**
1,2,3	1,3	2,4	4	All are
only	only	only	only	correct

292. The total O_2 in the blood will
  1. be most closely related to the hemoglobin content
  2. be most closely related to the pO_2 of the blood
  3. be reduced in anemia
  4. not be reduced in hypoxia

293. The compliance of a lung that changes in volume 1 L when the intrapleural pressure is lowered by 5 cm of H_2O would be
  1. calculated by dividing $\Delta P/\Delta V$
  2. calculated by dividing $\Delta V/\Delta P$
  3. 5.0 L/cm of H_2O
  4. 0.20 L/cm of H_2O

294. Perfusion without ventilation
  1. acts like an arteriovenous shunt
  2. acts like an increased partial pressure of O_2 in the alveoli
  3. leads to a depletion of O_2 in the alveoli involved
  4. acts like an increase in dead space

295. The ventilation perfusion ratio (V/Q)
  1. is determined by dividing cardiac output by alveolar ventilation
  2. in a patient with cardiac output of 12 L/min and alveolar ventilation of 2 L/min would be equal to 6
  3. is a constant for a given individual
  4. is determined by dividing the total alveolar ventilation by the cardiac output

296. The pontine respiratory center has
1. afferents directly to motoneurons necessary for respiration
2. an area that will cause prolonged inspiration if stimulated
3. its effects through a cortical-medullary reflex arc
4. an area that is primarily active in controlling the rate of respiration

297. When a patient has an extremely low arterial pO_2
1. the CNS will be directly stimulated
2. the presentation of 100% O_2 can be catastrophic
3. normal respiration cannot occur
4. the carotid body is most responsible for sensing the condition

298. Chemoreceptor drive constitutes an important compensatory mechanism in
1. carbon monoxide poisoning
2. anemia
3. methemoglobinemia
4. emphysema

299. The work of breathing is made up in part of work
1. required to overcome inertia of tissues
2. required to stretch elastic elements in the chest
3. necessary to overcome airway resistance
4. necessary to hold the bronchi open

300. Dyspnea is
1. not normal resting breathing
2. not increased breath rate and depth
3. the sensation of inadequate or distressful breathing
4. normal resting breathing

301. In Figure 12 the greatest change in volume during expiration and inspiration would occur at site(s)
1. A, B, and C
2. B and D
3. C and E
4. D only

Figure 12.

Directions Summarized				
A	**B**	**C**	**D**	**E**
1,2,3	1,3	2,4	4	All are
only	only	only	only	correct

302. In Figure 13
   1. the somewhat S-shaped curves are due to the effect of CO_2 on hemoglobin
   2. curve X is at a higher pH than curve Y
   3. curve Y is at a higher pH than curve X
   4. the somewhat S-shaped curves are due to a changing affinity of O_2 by hemoglobin at increasing pO_2

303. In Figure 13 a shift from curve Y to curve Z would occur if peripheral tissues developed a higher
   1. temperature
   2. 2,3-diphosphoglycerate
   3. CO_2
   4. myoglobin content

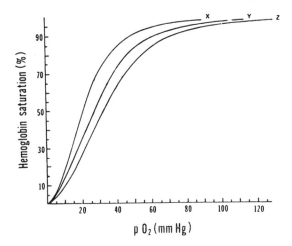

Figure 13.

304. The diaphragm
    1. is innervated by the vagus nerve
    2. decreases the volume of the thoracic space when it contracts
    3. increases intrathoracic pressure when it contracts
    4. decreases intrathoracic pressure when it contracts

305. Lung surfactant
    1. facilitates O_2 diffusion through alveolar membranes
    2. facilitates CO_2 diffusion through alveolar membranes
    3. increases surface tension of the alveolar membrane
    4. decreases the likelihood of alveolar collapse during expiration

306. Within the lungs the molecules of any gas
    1. are in continuous motion
    2. collide with other molecules that rebound, which results in a "pressure" development
    3. that develop a pressure will have a pressure whose magnitude is determined by the number of molecules present
    4. do not move randomly nor can they develop a pressure

307. There are several factors controlling adult human respiration on a moment-to-moment basis. Those which are of importance include:
    1. pulmonary stretch reflexes
    2. systemic arterial pCO_2 on carotid and aortic chemoreceptors
    3. pCO_2 of CNS capillary blood on chemosensors of the medulla
    4. cerebrospinal fluid pH

DIRECTIONS (Questions 308–320): Each of the questions or incomplete statements below is followed by five suggested answers or completions. Select the **one** that is **best** in each case.

---

308. A patient has a long-standing problem of diabetes mellitus (insulin deficiency) which you have been able to stabilize with daily insulin therapy. Suddenly this patient goes into a condition of severe acidosis and is brought into the hospital. You should EXPECT which of the following laboratory findings?
    A. A decrease in plasma pH, an increase in plasma HCO_3^-, and an acidic urine
    B. A decrease in plasma HCO_3^-, an increase in plasma pH, and an acid urine
    C. A urine that is acidic, a decrease in plasma HCO_3^-, and a decrease in plasma pH
    D. A urine that is alkaline, a decrease in plasma HCO_3^-, and a decrease in plasma pH
    E. Depressed respiration

309. In severe exercise the oxygen debt is
    A. surplus O_2 borrowed from the inspiratory reserve during exercise
    B. the average O_2 used less the resting O_2 for a similar time; required for anaerobic processes
    C. O_2 used above resting levels borrowed to provide the increased aerobic oxidations
    D. O_2 used after exercise above resting level and representing in part the anaerobic contribution in work
    E. decreased

310. Which of the following sets of values is indicative of compensated metabolic alkalosis?
A. $HCO_3^- = 20$ mEq/L, $pCO_2 = 25$ mm Hg, pH = 7.5
B. $HCO_3^- = 42$ mEq/L, $pCO_2 = 45$ mm Hg, pH = 7.5
C. $HCO_3^- = 17$ mEq/L, $pCO_2 = 30$ mm Hg, pH = 7.3
D. $HCO_3^- = 34$ mEq/L, $pCO_2 = 10$ mm Hg, pH = 7.7
E. $HCO_3^- = 17$ mEq/L, $pCO_2 = 19$ mm Hg, pH = 7.9

311. You are examining a patient who has an arterial pCO_2 of 40 mm Hg. During a test period the pCO_2 of expired gas is found to be 20 mm Hg and a respiratory minute volume of 8 L is noted. From this data the
A. alveolar ventilation is 8 L/min
B. alveolar ventilation is 40 L/min
C. alveolar ventilation is 4 L/min
D. tidal volume is 8 L
E. test should be stopped before potentially fatal cardiac arrhythmias occur

312. Respiratory alkalosis is characterized by
A. low pH concentration
B. fall in pCO_2
C. excess pulmonary ventilation
D. protein synthesis increases
E. reduced HCO_3^-/pCO_2 rate

313. Dynamic collapse of airways occurs
A. during vigorous inspiration
B. mostly in the respiratory bronchioles where walls are thin
C. in major bronchi during violent expiration because these airways have lower internal pressures than do smaller airways further upstream (nearer the alveoli)
D. only in unusual disease conditions
E. during normal quiet expiration

314. The primary stimulus of respiration is a
  A. twofold increase in the pCO_2 of inspired air
  B. twofold increase in the pO_2 of inspired air
  C. 50% decrease in the pCO_2 of inspired air
  D. 50% increase in the pO_2 of inspired air
  E. 50% decrease in pCO_2

315. The periodic nature of normal respiration is fundamentally caused by
  A. intermittent bursts of activity from cells in the pontine "apneustic center"
  B. feedback loops involving the peripheral chemoceptors
  C. coupled oscillatory behavior of inspiratory and expiratory cells in the medulla
  D. "pneumotaxic center"
  E. conscious control from areas of the motor cortex

316. The limitation of the pulmonary diffusion process
  A. is no movement of the molecules through the alveolar gas phase
  B. involves CO_2 equilibrium across the alveolar membrane
  C. prevents normal individuals from ever reaching alveolar-capillary pO_2 equilibrium
  D. is determined by reaction kinetics of the O_2 hemoglobin association and by O_2 solubility
  E. is largely determined by the diffusion capacity of the alveolar capillary membrane and capillary plasma

317. You are presented with a patient that you suspect has a compensated metabolic acidosis. Which of the following sets of lab data would confirm your suspicion?
  A. $HCO_3^- = 17$ mEq/L, $pCO_2 = 19$ mm Hg, pH = 7.9
  B. $HCO_3^- = 34$ mEq/L, $pCO_2 = 10$ mm Hg, pH = 7.7
  C. $HCO_3^- = 17$ mEq/L, $pCO_2 = 30$ mm Hg, pH = 7.3
  D. $HCO_3^- = 24$ mEq/L, $pCO_2 = 45$ mm Hg, pH = 7.5
  E. $HCO_3^- = 20$ mEq/L, $pCO_2 = 25$ mm Hg, pH = 7.5

318. Breathing CO_2 for prolonged periods results in
A. only alveolar pCO_2 rises
B. only tissue pCO_2 decreases
C. alveolar ventilation decreases
D. both alveolar and tissue pCO_2 increase
E. liver cirrhosis

319. In contrast to the systemic circulation, the pulmonary circulation is characterized by
A. low mean pressure
B. high resistance
C. relative small pulse pressure
D. large volume flow per minute
E. absence of sympathetic control

320. The total quantity of air that can be expelled from the lungs following a maximal inspiration is known as the
A. vital capacity
B. tidal volume
C. expiratory reserve volume
D. functional residual capacity
E. total capacity

DIRECTIONS (Questions 321–323): This section consists of a case history, followed by a series of questions. Study the history and select the one **best** answer to each question following it.

Case History (Questions 321–323):
A 50-year-old woman enters the hospital because of cough, weight loss, and dyspnea. She was found to have a mass in her right middle lobe bronchus. On radiographic analysis the area of the right middle lobe was denser than usual and somewhat smaller in size. The following values were obtained:

	Patient	*Normal*
Vital capacity	4.5	6 L
FRC (functional reserve capacity)	3.0	1.7 L
$FEV_{1.0}$ (forced expiratory volume)	2.0	4.5 L
MBC (maximum breathing capacity)	90	165 L/min
PaO_2	65	90 mm Hg
$PaCO_2$	43	38–42 mm Hg
PaO_2 during O_2 breathing	160 mm Hg	550 mm Hg
$PaCO_2$ during O_2 breathing	43 mm Hg	38–42 mm Hg

A single-breath N_2 washout test showed a progressive rise of nitrogen concentration over the course of expiration at a slope greater than normal. There was an abrupt rise in expired N_2 concentration which occurred at the point where the patient had exhaled to functional residual capacity (FRC).

To evaluate for possible surgery, a balloon-tipped catheter was placed in the right pulmonary artery (PA) and the artery was occluded by balloon inflation. The following data were obtained:

Mean PA pressure	40 mm Hg
Cardiac output	4 L/min
Total pulmonary diffusion capacity	15 mL O_2/min/mm Hg

321. In this patient
    A. the arterial hypoxemia present during air breathing is primarily caused by generalized hypoventilation
    B. there is no evidence which would support the existence of an abnormally large anatomic shunt
    C. an increased compliance alone could cause the abnormal $FEV_{1.0}$, FRC, N_2 washout
    D. the vessels of the left lung are probably normal
    E. a decreased compliance could explain all of the above data

322. The above data indicate
     A. the existence of a significant physiologic dead space
     B. the effective peribronchial pressure during expiration is probably lower than normal in this patient
     C. impaired diffusion does not become manifest as arterial blood hypoxemia while exercising this patient when she has had one pulmonary artery occluded
     D. a reduced ventilation/perfusion (V/P) ratio in the right middle lobe would appear as an increased physiologic shunt measured while breathing room air
     E. an anatomic shunt since this patient has a greatly reduced V/P

323. The data from this patient indicate that
     A. collapse of the right middle lobe behind an obstructing tumor can cause all the abnormalities except the N_2 washout and the reduced FRC
     B. blood flow through the partially obstructed middle lobe is reduced by the effect of alveolar hypoxia acting on the vessels. This is opposite to the effect of alveolar hypoxia in the fetal lung
     C. calculation of the physiologic shunt fraction requires knowledge of mixed systemic venous blood O_2 content in addition to the data given above
     D. the N_2 washout curve indicates that the distribution of inspired gas, or the sequence of regional emptying, or both, are as uniform as normal in this patient
     E. there is no abnormality of respiratory function

DIRECTIONS (Questions 324–337): Each group of questions below consists of a set of lettered components, followed by a list of numbered words or phrases. For each numbered word or phrase, select the **one** lettered component that is most closely associated with it. Each lettered component may be selected once, more than once, or not at all.

Questions 324–330:
  A. Anemic hypoxia
  B. Hypoxic hypoxia
  C. Histotoxic hypoxia
  D. Circulatory hypoxia

324. Breathing atmospheric air at high altitudes

325. Thromboembolism occluding a leg vein

326. Venous pO_2 higher than normal

327. Mercury poisoning of respiratory enzymes

328. Carbon monoxide poisoning of hemoglobin

329. Excessive blood loss

330. Drowning

Questions 331–337:
  A. Apneustic center
  B. Pneumotaxic center
  C. Medullary respiratory center
  D. Apneustic and pneumotaxic centers
  E. Apneustic and medullary centers

331. Required for rhythmic breathing

332. Role seems to be similar to that of vagal afferents

333. Present in the pons

334. Center for inspiratory drive

335. Center for expiratory drive

336. Stimulated by elevated serum pCO_2

337. Ablation results in apneustic breathing or slow deep breathing

Explanatory Answers

265. A. Inspiration is an active event. Expiration during quiet breathing is a passive event. There is a tendency for the chest wall to expand outward and the lung to move inward during relaxation. The latter event is due to the elastic recoil of the lung. These forces are opposite in direction but equal in magnitude. The respiratory activity is controlled by the medulla and pons with higher influences from the cerebral cortex, as when one wants to hold their breath. (REF. 3, pp. 570–575)

266. A. CO_2 is carried in the blood in combination with hemoglobin as dissolved CO_2 gas, and mainly as bicarbonate. (REF. 3, pp. 576–577)

267. E. Many things influence O_2 uptake and CO_2 removal. These include stretch receptors in the lung, chemoreceptors in the arteries, and changes in metabolic activity. Due to their relative positions, the pumping of the heart exerts a mechanical action on the lung and promotes the distribution of O_2 within the lungs. Intrathoracic pressure can influence venous return and consequently cardiac output. (REF. 3, p. 577)

268. E. The diaphragm is a dome-shaped muscle that flattens when contracted. This contraction increases the anteroposterior, lateral, and vertical dimensions of the thoracic cage. The positive pressure in the abdomen forces the lower ribs outward. (REF. 3, p. 578)

269. D. The diaphragm is the principle muscle of respiration, and accounts for over two thirds of inspired air, and even more during anesthesia. Two thirds of the fibers of the diaphragm are slow-twitch which help prevent the diaphragm from becoming fatigued. It is supplied by the phrenic nerve. (REF. 3, p. 578)

270. B. The energy expended during breathing can be measured in terms of oxygen cost. This is measured at rest during normal ventilation and at increased voluntary hyperventilation. Oxygen consumption of the breathing apparatus is usually 1 mL/L or less than 5% of total body consumption of O_2. (REF. 3, p. 579)

271. D. The functional residual capacity is the volume of air in the lungs at normal, end-expiratory, resting state. (REF. 3, p. 579)

272. B. Total volume is the amount of air inhaled and exhaled during breathing. The maximum volume that can be inhaled from functional residual capacity is the inspiratory capacity (IC). The volume of air left at the end of normal, resting exhilation is the functional residual capacity. The volume at maximum inspiration is total lung capacity. (REF. 3, p. 579)

273. A. Elastin and collagen give lung tissue its resilience. Elastin allows stretching while collagen prevents over-stretching. Emphysema is a disease that degrades elastin and collagen as well as the alveolar walls. The result is an increase in the distensibility of the lung. (REF. 3, p. 581)

274. E. Chest wall expansion represents 70% of total lung capacity if opposed by the lung. However, at values greater than 70% the chest wall recoils inward, at volumes less than 70% it recoils outward. (REF. 3, p. 503)

275. A. Flow rate is directly proportional to the driving pressure, varies inversely with the length of the tube and viscosity, and is to the fourth power of the tube radius. (REF. 3, pp. 584–585)

276. E. The parabolic profile of laminar flow is characteristic of both air and fluid. However, gas molecules can move laterally and collide to form eddys much more readily. Gas flow rate varies with the square root of the driving pressure. (REF. 3, pp. 585–586)

277. B. Resistance is very high for nasal passage but air flow decreases as cross-sectional area increases, which is what happens with each successive branching of the bronchi. Therefore, air flow resistance decreases with each successive branching. (REF. 3, p. 585)

278. D. At the beginning of forced expiration, just as the lung volume begins to decrease, lung elastic recoil goes down as well as the transmural pressure. The intrathoracic airways narrow which increases airway resistance. (REF. 3, p. 587)

279. B. During a maximal inspiratory effort, the pleural pressure is subatmospheric. The muscles of inspiration diminish in their force generation according to length tension properties of muscle. This results in a progressive reduction in air flow. The transmural pressure is large. (REF. 3, pp. 587–588)

280. A. Approximately 60% of the CO_2 in venous blood is in the form of bicarbonate, with 7% in the dissolved state, and 23% in the carbamino form. (REF. 1, p. 500)

281. E. All the factors listed shift the curve to the right and down, unbinding the O_2 with hemoglobin at any given alveolar pO_2. (REF. 1, p. 498)

282. D. Intrapulmonary pressure is determined by airway resistance and direction of airflow. (REF. 2, p. 991 ff)

283. A. A major role of respiration is the removal of CO_2. A low pCO_2 results in a decrease in ventilation. (REF. 2, p. 994 ff)

284. E. All the factors listed. The higher centers in the central nervous system have an influence for the respiratory increases. (REF. 1, p. 504 ff)

285. C. The diffusion coefficient of O_2 in tissues is 20 times lower than that for CO_2 because the diffusion coefficient for the transfer of each gas through the respiratory membrane depends on its solubility in the membrane and inversely on the square root of its molecular weight. (REF. 1, p. 483 ff)

286. E. Sensory input from the vagus nerve to the respiratory centers in the brain stem is necessary for normal rhythm rate and depth of respiration. (REF. 1, p. 504)

287. E. Severe gastric vomiting from deep in the GI system will result in a metabolic acidosis with CO_2 accumulation, and corresponding increases in bicarbonate; respiration will occur to compensate for the acidosis. (REF. 1, p. 450)

288. D. Although hemoglobin is important for CO_2 transport, the mechanism is important and completely different than that for O_2. The factors listed illustrate the importance of hemoglobin as a reserve for O_2. (REF. 1, p. 496)

289. E. The release of O_2 by hemoglobin in vascular beds is facilitated by low pO_2, low pH, high temperature, and high pCO_2. (REF. 1, p. 498)

290. C. Carbon monoxide is specific in that it has a very high affinity for a specific site on the hemoglobin molecule. (REF. 1, p. 500)

291. B. The Hering-Breuer reflexes result in inhibition of inspiration when the lungs are inflated and excitation of inspiration when the lungs are deflated. (REF. 1, p. 505)

292. B. The total O_2 in the blood will be largely determined by how much hemoglobin is present. Hence, when red cell numbers decrease in anemia, total O_2 goes down. (REF. 1, p. 48)

293. C. The compliance of a lung is calculated by dividing ΔV by ΔP. For a lung showing a volume change of 1 L with a pressure change of 5 cm of H_2O, the compliance would be 0.20. (REF. 1, p. 468)

294. B. If an alveolus has perfusion without ventilation, the O_2 in the alveolus will be depleted until it is at the level of mixed venous blood. At this point the blood flowing around the alveolus will appear as if it had passed through an arteriovenous shunt. (REF. 1, p. 490 ff)

295. D. The ventilation perfusion ratio is determined by dividing the total alveolar ventilation by the cardiac output. (REF. 1, p. 490 ff)

296. C. The pontine respiratory center contains the apneustic center, which, if stimulated, will cause apneustic breathing, and the pneumotaxic center, which is largely concerned with controlling the rate of respiration. (REF. 1, p. 504 ff)

297. C. The presentation of 100% O_2 to a patient with a prominent carotid body drive (i.e., a low arterial O_2) often results in respiratory arrest. (REF. 1, p. 524)

298. D. Pulmonary emphysema is a degenerative disorder leading to disruption of alveolar septa and pulmonary fibrosis with a decrease in efficiency of alveolar ventilation. In chronically and severely hypoxic persons, the slowly adapting chemoreceptors provide not only reflex activation of the respiratory centers, but an arousal mechanism for higher functions such as consciousness. (REF. 3, p. 632 ff)

299. A. The work of breathing can be broken up into a series of elements including the work against inertia, the work against elastic elements, and the work to overcome airway resistance. (REF. 1, p. 469 ff)

300. A. Dyspnea is defined as the sensation of inadequate or distressful breathing. (REF. 1, p. 522 ff)

301. D. Most of the volume change in the lung occurs in the respiratory bronchioles. (REF. 1, p. 486)

302. C. The curves shown in Figure 13 demonstrate several findings. Two important items are that at increasing pHs, there is an increased affinity of hemoglobin for O_2, and that the affinity of hemoglobin for O_2 changes as the pO_2 varies. (REF. 1, p. 498)

303. A. Higher temperature, concentration of 2,3-diphosphoglycerate, and pCO_2 all shift hemoglobin oxygen saturation curves to the right. (REF. 1, p. 498)

304. D. The major action of the diaphragm is to decrease the intrathoracic pressure when it contracts. (REF. 1, p. 446)

305. D. Lung surfactant is a phospholipid that adjusts the surface tension of the air liquid interface of the alveoli in order to decrease the likelihood of alveolar collapse during expiration. The surface tension of the alveoli is maintained at a relatively constant and low level of lung surfactant that is concentrated as the alveoli shrink during expiration. (REF. 1, p. 467 ff)

306. A. The activities of gas molecules in the lung are governed by the gas laws. (REF. 1, p. 481 ff)

307. E. The control systems for respiration include all the factors. The most sensitive factor appears to be the pCO_2 of the blood. (REF. 1, p. 507)

308. C. The response of the body to a low plasma pH, due to a metabolic imbalance is to decrease plasma HCO_3^- by blowing off CO_2 and to excrete increased amounts of H^+ in the urine. (REF. 1, p. 448 ff)

309. D. After exercise, humans continue to use oxygen at greater than rest levels. The difference between postexercise and resting levels of O_2 uptake is known as oxygen debt. (REF. 1, p. 843)

310. B. During compensated metabolic alkalosis there is a gain of CO_2 in an attempt to return the pH toward normal. The result is increased HCO_3^-, pCO_2, and a near-normal pH. (REF. 1, p. 446 ff)

311. C. Alveolar ventilation = Respiratory volume ×

$\dfrac{\text{Expired } pCO_2}{\text{Arterial } pCO_2}$. (REF. 1, p. 471)

312. B. During respiratory alkalosis pCO_2 decreases because CO_2 is blown off during excess ventilation with a concomitant fall in H^+ concentration. (REF. 1, p. 448)

313. C. Dynamic collapse of bronchioles occurs with forced expiration when the external pressure applied to the muscles of respiration exceeds the internal pressure of the bronchi. (REF. 1, p. 474 ff)

314. A. An increase in pCO_2 from the normal level of 40 mm Hg to 63 mm Hg causes a tenfold increase in alveolar ventilation. An increase in pCO_2 stimulates alveolar ventilation, not only directly but also indirectly through its effects on hydrogen ion concentration. (REF. 1, p. 506 ff)

315. C. Groups of neurons in the medulla which give rise to alternating inspiratory and expiratory activity are the fundamental source of respiratory rhythmicity. (REF. 1, p. 504 ff)

316. E. The red cell component of lung diffusion represents only 1% to 5% of the total lung resistance to diffusion of O_2. (REF. 1, p. 577 ff)

317. C. Increased pulmonary ventilation is seen in compensated metabolic acidosis. It is characterized by rapid elimination of CO_2 in an attempt to raise pH. The respiratory compensation is partial (50%–75%). (REF. 1, p. 449)

318. D. CO_2 equilibrates between serum and tissues very rapidly. (REF. 1, p. 485 ff)

319. A. The capillary pressure of 7 mm Hg is almost exactly halfway between the mean pulmonary arterial pressure of 13 mm Hg and the left atrial pressure of 2 mm Hg, indicating the arterial and venous resistances of the lungs are approximately equal. This is in marked contrast to the systemic circulation, in which the arterial resistance is four to seven times as great as the venous resistance. (REF. 1, p. 280 ff)

320. A. The vital capacity is the sum of the tidal volume and the inspiratory and expiratory reserve volumes. (REF. 1, p. 471)

321. C. The patient's vital capacity is not that low. An anatomic shunt would be evident if the ventilation/perfusion (V/P) ratio were greatly reduced, which it is not in this patient. In addition, the vessels of the lung are abnormal because the patient's cardiac output is normal. Hence, a great proportion of the output must be going to the lungs. However, an increased compliance can cause the abnormalities seen above. (REF. 1, p. 490 ff)

322. D. Although there is no real anatomic shunt (because the V/P ratio is not greatly reduced), the breathing of room air results in an apparent physiologic (not anatomic) shunt. (REF. 1, p. 490 ff)

323. C. The shunting of mixed venous pulmonary arterial blood around alveolar capillaries and directly into the pulmonary veins can be detected by a 100% oxygen breathing test because the level of oxygenation of mixed venous blood would be directly proportional to the amount of blood that is shunted. (REF. 1, p. 490)

324. B. Atmospheric air at high altitudes has low partial pressures of all gases, hence the pO_2 is lowered sufficiently to give hypoxic hypoxia. (REF. 1, p. 522)

325. D. The reduction in blood flow due to a thromboembolism occluding a leg vein will give rise to stagnantoxia or circulatory hypoxia. (REF. 1, p. 522)

326. C. A venus pO_2 higher than normal would indicate that the perfused tissue is unable to utilize the O_2 being presented to it. This is usually classified as histotoxic hypoxia. (REF. 1, p. 522)

327. C. If the respiratory enzymes of a tissue are poisoned by mercury, they will be unable to utilize O_2 and a histotoxic hypoxia will result. (REF. 1, p. 522)

328. A. Carbon monoxide poisoning removes some of the hemoglobin available for O_2 transport. This results in an anemic anoxia. (REF. 1, p. 522)

329. A. Excessive blood loss also removes some of the hemoglobin available for transport and results in an anemic anoxia. (REF. 1, p. 522)

330. B. Immersion of the head in water cuts off the supply of fresh air to the lungs and results in an hypoxic hypoxia. (REF. 1, p. 522)

331. C. The medullary respiratory center is responsible for alternating inspiration that is the essential activity of breathing. (REF. 1, p. 504 ff)

332. B. The role of the pneumotaxic center is to control the rate of respiration, and is similar in action to the vagal afferents. (REF. 1, p. 504 ff)

333. D. Both the apneustic and pneumotaxic centers are found in the pons. (REF. 1, p. 504 ff)

334. E. The apneustic and medullary centers supply the drive for inspiration. (REF. 1, p. 504 ff)

335. C. The medullary respiratory center is responsible for supplying the drive for expiration. (REF. 1, p. 504 ff)

336. E. Both the apneustic center and the medullary centers are activated by elevated serum pCO_2. (REF. 1, p. 504 ff)

337. B. Ablation of the pneumotaxic center releases the apneustic center from its influence. This results in apneustic or slow deep breathing. (REF. 1, p. 504 ff)

5 Renal Physiology

DIRECTIONS (Questions 338–371): For each of the questions or incomplete statements below, **one** or **more** of the answers or completions given is correct. Select

   A if only *1, 2, and 3* are correct
   B if only *1 and 3* are correct
   C if only *2 and 4* are correct
   D if only *4* is correct
   E if all are correct

338. The role of the kidney in homeostasis may include which of the following?
    1. Regulation of extracellular fluid composition
    2. Regulation of red blood cell formation
    3. Secretion of certain hormones, such as angiotensin II, prostaglandins, and kinins in the regulation of blood pressure
    4. Secretion of erythropoietin

339. Body fluid compartments
    1. account for less than 33% of total body water
    2. fluctuate in size in response to environmental stress or disease
    3. do not include bone or cartilage matrices
    4. accounts for 60% of total body mass

125

		Directions Summarized		
A	**B**	**C**	**D**	**E**
1,2,3	1,3	2,4	4	All are
only	only	only	only	correct

340. Which of the following is(are) true concerning the various fluid compartments of the body?
  1. In a steady-state condition, osmotic activity is nearly the same for all body compartments
  2. Ions and their distribution can affect the volume of fluid compartments
  3. Movement of water (loss or gain) is in proportion to the volumes of each compartment
  4. Total body water and the fluid volumes of each compartment change very little from day to day

341. The flow of fluid through the kidney
  1. requires an arterial blood supply via the afferent arteriole
  2. begins by entering a filtering unit called Bowman's capsule
  3. passes through the ducts of Bellini before exiting from the kidney
  4. begins in the proximal convoluted tubule by filtration, followed by reabsorption and secretion of the filtrate

342. Which is a feature of the glomerulus?
  1. Podocyte
  2. Fenestrated epithelium
  3. Slit membrane
  4. Foot processes

343. Which of the following is(are) true regarding the Starling hypothesis for capillary exchange?
    1. Requires knowing the protein concentration in the proximal tubular fluid
    2. Describes how fast fluid crosses the glomerulus under a given filtration pressure
    3. Is the product of coefficient K times the algebraic sum of the glomerular capillary hydrostatic, proximal tubular hydrostatic, glomerular capillary oncotic, and proximal tubular fluid oncotic pressures
    4. Can only be estimated for the kidney as a whole

344. Which of the following is true regarding glomerular filtration rate (GFR)-capillary flow?
    1. When flow is decreased, the plasma oncotic pressure increases more over a given distance along the capillary
    2. GFR remains relatively constant because the mean hydrostatic capillary pressure remains unaltered
    3. At higher capillary flow rates, the plasma oncotic pressure is lower, net filtration pressure is elevated, and GFR increases
    4. At higher capillary flow rates, the net filtration pressure is reduced, thereby lowering the GFR

345. Which of the following does not increase the filtration coefficient K?
    1. Angiotensin II
    2. Acetylcholine (a vasodilator)
    3. Norepinephrine (a vasoconstrictor)
    4. Prostaglandins

346. Which of the following classes of molecules are transported by the renal epithelia?
    1. Aliphatic acids
    2. Aromatic acids
    3. Organic bases
    4. Monovalent and polyvalent ions

		Directions Summarized		
A	**B**	**C**	**D**	**E**
1,2,3	1,3	2,4	4	All are
only	only	only	only	correct

347. Processes involved in the transport of materials across the tubular epithelium are dependent upon
    1. size of the molecules
    2. concentration gradient
    3. charge
    4. lipid solubility

348. Characteristics of active transport systems include:
    1. molecular specificity
    2. saturation at low concentrations
    3. transport rate higher than would normally be predicted for a given lipid solubility and size of a molecule
    4. insensitivity to inhibitors

349. Plasma threshold of a substance
    1. means that over a small range of plasma concentrations excretion varies linearly
    2. is descriptive of molecules that display tubular transport maximum (T_m) type characteristics
    3. is descriptive of a gradient-time system
    4. is the minimal plasma concentration at which excretion begins

350. Gradient-time system characteristics include:
    1. no evidence of plasma threshold
    2. excretion varies linearly
    3. reabsorption rate varies continuously with plasma concentration
    4. excretion remains constant over a wide range of filtration rates

351. Renal function can be tested by infusing a secreted dye at progressively faster rates until saturation is achieved. Analyzing urine under these conditions
   1. yields information on tubular function
   2. yields information on cardiovascular function
   3. yields information on cardiovascular as well as tubular function
   4. yields information on tubular function and tubular transport capacity

352. According to the equation for net tubular transport rate, TR = filtration rate − excretion rate, if the transport rate is positive
   1. excretion rate is in excess of filtration rate
   2. secretion must have taken place
   3. solute must have been added to the glomerular filtrate
   4. reabsorption must have taken place

353. The loops of Henle of the outer cortical nephrons
   1. do not play any important role in overall renal function, and are simply unimportant vestiges of evolutionary development
   2. do not participate in the urinary diluting mechanism
   3. are functionally unimportant in the renal conservation of sodium and water
   4. do not contribute to the medullary osmotic gradient

354. In active transport there must be
   1. binding of solute to some membrane component
   2. presence of carrier molecule
   3. energy required for transport
   4. directional sites in membrane pores

355. Under normal conditions in humans, the colloid osmotic pressure of the blood at the capillary level
   1. is balanced by capillary hydrostatic pressure
   2. is balanced by capillary oncotic pressure
   3. is due primarily to presence of nondiffusible proteins
   4. tends to inhibit the diffusion of appropriate substances for the nutritive needs of cells

		Directions Summarized		
A	**B**	**C**	**D**	**E**
1,2,3	1,3	2,4	4	All are
only	only	only	only	correct

356. For those substances that are actively reabsorbed, the maximal amount that can be transported per unit time by the kidney tubules
   1. is termed the tubular transport maximum
   2. requires specific transport systems for each substance transported
   3. depends on the maximum rate at which the transport mechanism itself operates
   4. is dependent upon tubular load

357. Clearance ratios greater than or equal to 1 are most likely seen with substances which are
   1. reabsorbed
   2. secreted
   3. bound to tubular proteins
   4. neither secreted nor absorbed

358. Solute particles move from the plasma of the renal glomerulus to the fluid in the Bowman's capsule by
   1. active transport
   2. diffusion
   3. renal flow
   4. bulk flow

359. NH_3 (ammonia) produced by the kidneys comes mainly from
   1. leucine
   2. glycine
   3. alanine
   4. glutamine

360. Polyuria (diuresis), which occurs in the diabetic with a GFR = 120 mL/min and blood sugar level = 350 mg%, is indicative of

1. diuresis due to reduced active transport of sodium out of the tubule because of diminished activity of the sodium pump
2. losses of water and sodium which can be prevented by administration of antidiuretic hormone (ADH) and an aldosteronelike mineralocorticoid
3. a cellular and extracellular hydration due to water retention of the glucose; hence diuresis is *never* observed in a diabetic individual
4. an osmotic diuresis due to glucosuria, and the water loss will exceed "salt" loss

361. The "renal plasma threshold" for glucose will be decreased by

1. an increase in glomerular filtration rate
2. a decrease in glucose T_m
3. a decrease in the slope of the glucose reabsorbtion curve
4. a decrease in tubular reabsorption

362. The tonicity of the urine as it enters the renal collecting duct may be

1. hypotonic
2. isotonic
3. hypertonic
4. hypotonic or isotonic but never hypertonic

363. Extracellular dehydration results in

1. stimulation of the volume and osmoreceptors and increased ADH secretion
2. inhibition of the volume and osmoreceptors and increased ADH secretion
3. increased extracellular osmolality
4. inhibition of the volume and osmoreceptors and decreased ADH secretion

Directions Summarized				
A	**B**	**C**	**D**	**E**
1,2,3	1,3	2,4	4	All are
only	only	only	only	correct

364. Of the following, which are CORRECTLY defined or described?
    1. Tubular maximum secretion: has a finite upper limit though exhibits a phenomenon analogous to the threshold phenomenon for reabsorption
    2. Clearance ratio: renal clearance of one substance divided by the clearance of another substance
    3. Filtration fraction: glomerular filtration rate divided by renal plasma flow
    4. Effective renal plasma flow: volume of plasma flow supplied to the entire renal tissue

365. The juxtaglomerular apparatus
    1. is involved in maintaining the normal balance of sodium in the body
    2. includes a long loop of Henle that dips into the medulla
    3. participates in the control of aldosterone secretion through the renin-angiotensin system
    4. functions as a sphincter around the distal tubules

366. About 4 to 6 days after you place a "normal" patient on a low sodium diet to reduce his or her weight, which of the following will be observed?
    1. Plasma renin and aldosterone are above normal
    2. Plasma renin and aldosterone are below normal
    3. Plasma sodium concentration is normal
    4. Plasma sodium is below normal

367. You have two patients (each weighing 80 kg), one of whom
you give 1000 mL of distilled water (subject A), while the
other you give 1000 mL of isotonic saline (subject B). Both
drink the fluid within the same length of time. Correct
statements regarding subjects A and B include:
   1. subject A has a greater change in plasma osmolality
   2. subject B has the greater increase in plasma volume
   3. subject A will have a greater urinary output within 2
      hours after the fluid intake
   4. subject B has the greater change in urine osmolality

368. The substance(s) which makes up the greatest part of the
reabsorptive "load" in the renal tubule is
   1. glucose
   2. urea
   3. potassium
   4. sodium

369. The renal "countercurrent" mechanism is dependent upon
the anatomic arrangement of the
   1. vasa recta
   2. collecting ducts
   3. loop of Henle
   4. proximal tubule

370. Secretion of aldosterone will result from
   1. low extracellular sodium
   2. low extracellular volume
   3. high extracellular potassium
   4. smoking, agitation, and stress

371. Aldosterone secretion is controlled by levels of
   1. angiotensin II
   2. ADH
   3. plasma volume
   4. plasma bicarbonate

DIRECTIONS (Questions 372–378): Each of the questions or incomplete statements below is followed by five suggested answers or completions. Select the **one** that is **best** in each case.

372. The volume of plasma needed each minute to supply a substance at the rate at which it is excreted in the urine is known as the
    A. diffusion constant of the substance
    B. clearance of the substance
    C. extraction ratio of the substance
    D. tubular mass of the substance
    E. filtration rate of the substance

373. Total renal blood flow of both human kidneys is what fraction of the resting cardiac output?
    A. 5%
    B. 10%
    C. 25%
    D. 40%
    E. 50%

374. An increase in the osmolality of the extracellular compartment will
    A. stimulate the volume and osmoreceptors, and inhibit ADH secretion
    B. inhibit the volume and osmoreceptors, and stimulate ADH secretion
    C. inhibit the volume and osmoreceptors, and inhibit ADH secretion
    D. stimulate the volume and osmoreceptors, and stimulate ADH secretion
    E. cause no change in ADH secretion

375. Which of the following combinations of data would lead you to suspect that a patient had the "syndrome of inappropriate antidiuretic hormone secretion" (SIADH)?

	P_{Osm} mOsm/kg	P_{Na} mEq/L	U_{mOsm} mOsm/kg
A.	286	138	627
B.	263	126	52
C.	286	138	177
D.	263	126	426
E.	300	144	100

376. If the renal plasma flow is 600 mL plasma/min, and the hematocrit is 40%, what is the renal blood flow (in mL/min)?
A. 1500
B. 1000
C. 960
D. 1200
E. 1800

377. The normal human glomerular filtration rate (GFR) is approximately (in mL/min)
A. 25
B. 50
C. 125
D. 300
E. 500

378. Figure 14 graphically represents the data collected from a patient following the oral ingestion of a solution. The solution probably was
A. 5% NaCl
B. 0.9% NaCl
C. 0.9% KCl
D. water
E. 15% glucose

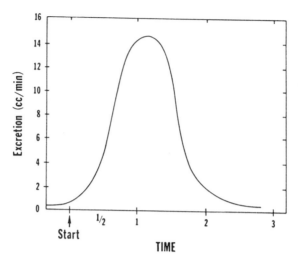

Figure 14.

DIRECTIONS (Questions 379–381): This section consists of a situation, followed by a series of questions. Study the situation, and select the **one** best answer to each question following it.

The following data were obtained in an 80-kg male patient during renal clearance tests:

Para-aminohippuric acid (PAH) concentration in plasma,	0.3 μg/mL
PAH concentration in urine,	90.0 μg/mL
Inulin concentration in plasma,	10.0 μg/mL
Inulin concentration in urine,	0.6 mg/mL
pO_4 concentration in plasma,	0.5 μM/mL
pO_4 concentration in urine,	1.0 μM/mL
Hematocrit,	40%
Urine flow,	2 mL/min

379. The glomerular filtration rate in mL/min is
   A. 120
   B. 150
   C. 180
   D. 240
   E. 400

380. The renal plasma flow in mL/min is
   A.   100
   B.   300
   C.   600
   D.   900
   E.   1200

381. The renal blood flow in mL/min is
   A.   100
   B.   300
   C.   600
   D.   1000
   E.   1200

DIRECTIONS (Questions 382–414): Each group of questions below consists of a set of lettered components, followed by a list of numbered words or phrases. For **each** numbered word or phrase, select the **one** lettered component that is most closely associated with it. Each lettered component may be selected once, more than once, or not at all.

Questions 382–387 (Figure 15):

   A. Site A   D. Site D
   B. Site B   E. Site E
   C. Site C

In Figure 15 the site(s) at which there is the greatest or highest

382. net fluid transport is

383. dilution of solutes is

384. amino acid reabsorption is

385. Na$^+$ reabsorption is

386. concentration of solutes is

387. active Na$^+$/Cl$^-$ transport is

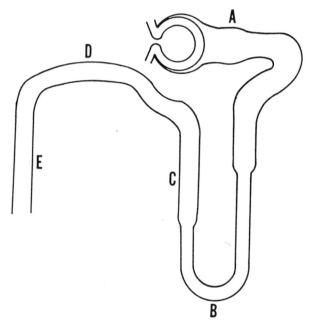

Figure 15.

Questions 388–395:

  A. Increases from the normal average value
  B. Decreases from the normal average value
  C. No change from the normal average value

A patient develops an increase in mean systemic blood pressure and blood volume. Indicate how the increased blood pressure and volume would affect the following:

388. Secretion of the juxtaglomerular cells of the kidney

389. Angiotensin secretion

390. Adrenal cortex stimulation

391. Aldosterone secretion

392. ADH secretion

393. Na^+ reabsorption

394. K^+ reabsorption

395. H_2O reabsorption

Questions 396–398:

    A. Increases from the normal average value
    B. Decreases from the normal average value
    C. No change from the normal average value

During the early hypotensive phase of hemorrhagic shock, indicate how the alterations with this early shock affect the following:

396. The hematocrit

397. The cardiac output

398. Peripheral resistance

Questions 399–401:

    A. Increases from the normal average value
    B. Decreases from the normal average value
    C. No change from the normal average value

During the irreversible phase of hemorrhagic shock

399. Plasma protein concentration

400. Precapillary to postcapillary resistance ratio

401. Mean arterial blood pressure

Questions 402–406:

    A. 4%–5% of lean body mass
    B. 12%–15% of lean body mass
    C. 20%–24% of lean body mass
    D. 40%–50% of lean body mass
    E. 60%–70% of lean body mass

402. Total body water

403. Interstitial compartment water

404. Intracellular water content

405. Vascular (plasma) water content

405. Extracellular fluid

Questions 407–414:

    A. Increases from the normal average value
    B. Decreases from the normal average value
    C. No change from the normal average value

407. pCO_2 of the plasma during respiratory acidosis

408. HCO_3^- concentration of the plasma during respiratory acidosis

409. pCO_2 of the plasma during respiratory alkalosis

410. HCO_3^- concentration of the plasma during respiratory alkalosis

411. pCO_2 of the plasma during metabolic acidosis

412. HCO_3^- concentration of the plasma during metabolic acidosis

413. pCO_2 of the plasma during metabolic alkalosis

414. HCO_3^- concentration of the plasma during metabolic alkalosis

Explanatory Answers

338. E. The function of the kidney is to regulate the extracellular fluid composition thereby providing a constant environment for the cells of the body to exist. The role of the kidney in the regulation of extracellular fluids and their composition is achieved through neural and humoral means. These mechanisms include secretion of angiotensin II, prostaglandins, and kinins, hormones essential to the regulation of blood pressure, as well as filtration, reabsorption, and secretion of the various components of the body fluids. Erythropoietin, another hormone secreted by the kidney, helps regulate red cell formation. (REF. 3, p. 745)

339. C. Total body water volume is separated into several compartments, the principle compartments being extracellular, intracellular, and interstitial. Sixty percent of total body weight can be attributed to the water content of the body. This includes water which is inaccessible, i.e., bound up in cartilage, tendon, and the matrices of bone (approximately 10%). This latter fluid compartment is termed the transcellular compartment. (REF. 3, pp. 745–746)

340. A. Fluid volumes as well as total body water can fluctuate greatly over the course of a day. However, osmotic activity of each compartment remains relatively balanced due to the proportional movement of fluid from each compartment. The Donnan effect explains the effect ion distribution has on fluid compartment size. (REF. 3, p. 748)

341. A. The flow of fluid through the kidney begins with the primary filtration in the glomerulus inside Bowman's capsule (not in the proximal tubules). The blood is supplied to the glomerulus via the afferent arteriol, passes through the glomerular capillaries, and exits through the efferent arteriole. This filtrate flows through a highly convoluted section called the proximal convoluted tubule. The tubule extends toward or into the medulla forming a hairpin turn called the loop of Henle with its descending and ascending loops. From the ascending loop fluid enters the distal convoluted tubule and from there into collecting ducts. The urine exits the kidney through small pores called the ducts of Bellini. (REF. 3, p. 749)

342. E. The podocytes are specialized epithelial cells which are a continuation of the proximal tubule endothelium. The fenestrated endothelium is a continuation of the arterioles supplying the glomerulus. The slit membrane spans two podocytes and acts as a molecular sieve. The foot processes are extensions of the podocytes. (REF. 3, pp. 753–754)

343. B. Knowing the protein concentration in the proximal tubular fluid allows one to apply the Starling hypothesis for capillary exchange. The algebraic expression is the product of the coefficient K times the sum of the glomerular capillary hydrostatic, proximal tubular hydrostatic, glomerular capillary oncotic, and proximal tubular fluid oncotic pressures. K is a coefficient that describes how fast fluid crosses the glomerulus under a given pressure. Values for K can be estimated for a single glomerulus or for an entire kidney. (REF. 3, pp. 752–753)

344. B. Plasma oncotic pressure rises significantly along the length of the glomerular capillary relative to the capillary flow rate. At lower capillary flow rates, the plasma spends more time at a particular location. The result is that the plasma oncotic pressure has a chance to rise more for a given distance traveling along the capillary. Therefore, at higher flow rates the oncotic pressure does not increase as quickly, and the hydrostatic pressure is higher, which increases the GFR. (REF. 3, p. 755)

345. E. Angiotensin II, bradykinin, and prostaglandins all decrease the filtration coefficient. Curiously, vasodilators, such as acetylcholine, and vasoconstrictors, such as norepinephrine, also decrease the filtration coefficient for reasons that are not yet clear. (REF. 3, p. 756)

346. E. Typical classes of molecules transported by the renal epithelia include monovalent and polyvalent ions such as Na^+, Cl^-, and PO_43^- and calcium, respectively. Aromatic and aliphatic acids such as penicillin and pyruvate, respectively, and organic bases such as thiamine are also transported. Large molecules such as glucose require a special transport system. (REF. 3, pp. 757–758)

347. E. All are intrinsic characteristics of the material being transported which affects the transportability of that material. In other words, size, charge, solubility, and concentration are factors that affect transport. Small uncharged molecules traverse rather easily. Large molecules traverse more slowly or require special transport mechanisms. (REF. 3, p. 758)

348. B. Active transport follows, to some extent, the kinetics of enzyme activity. Active transport systems are sensitive to inhibitors and compete for similar molecules despite their molecular specificity. They become saturated at high concentrations. (REF. 3, p. 758)

349. C. Using glucose, which displays T_m characteristics, as an example, increasing plasma glucose concentration from zero produces no change in excretion initially. However, once glucose reaches a plasma level of about 200 mg/dL of plasma, excretion begins to rise curvilinearly. This curvilinear range is termed the splay. Concentrations above the splay imply that the system is saturated, the difference between filtration and reabsorption is a constant, and any increase in filtration directly increases excretion. (REF. 3, p. 760)

350. A. It is assumed that the gradient-time system has a lower affinity relative to plasma concentrations. Therefore, it does not become saturated. Using sodium as an example, 67% of the sodium filtered is reabsorbed over a wide range of filtration rates. Since excretion varies linearly over a wide range of plasma concentrations, there is no evidence of a threshold. (REF. 3, p. 760)

351. E. Analysis of urine under saturation conditions gives an estimate of the tubular transport capacity. This transport capacity also depends on renal blood flow. As a result, this type of testing yields information on cardiovascular as well as tubular function. (REF. 3, p. 761)

352. D. According to the equation, net transport rate = filtration rate − excretion rate, if the transport rate is negative than excretion rate is greater than filtration rate. In other words some secretion must have taken place (solute has been added to glomerular filtrate). On the other hand, if less material appears in the final urine than originally had been filtered, then reabsorption has taken place. (REF. 3, pp. 760–761)

353. D. The countercurrent mechanism responsible for the secretion of hyperosmotic urine requires the penetration of loops of Henle into the renal medulla for the development of a medullary osmotic gradient. The loops of the outer cortical nephron do not descend into the inner medulla. (REF. 3, p. 787 ff)

354. A. An active transport system must have energy, a carrier system, and some method for sequestering the transported system when it reaches the cell membrane. (REF. 2, p. 25 ff)

355. B. The functional capillary pressure is about 17 mm Hg while the tissue pressure is nearly 0. The presence of nondiffusible (at best poorly diffusible) proteins in the capillaries exerts an osmotic pressure equal but opposite to the capillary hydrostatic pressure. This osmotic pressure due to the presence of these proteins is called the colloid osmotic or oncotic pressure. (REF. 1, p. 221 ff)

356. A. The tubular transport maximum (T_m) is the maximal amount of a given material that can be transported across the tubular membrane per unit time. This transport maximum depends largely on the speed at which the transport mechanism itself can operate. Since substances are reabsorbed or secreted, T_m is independent of tubular load, which is the amount of substance filtered through the glomerulus each minute. However, the relationship between tubular load and T_m will affect final urine concentration of the substance. (REF. 1, p. 408 ff)

357. C. The ratio of the amount of a substance filtered to the amount of that same substance cleared in the urine is the clearance ratio for that substance. When secretion is involved one sees clearance ratios greater than 1. With reabsorption, clearance ratios are less than 1. For substances that are neither secreted nor absorbed, clearance = GFR. (REF. 2, p. 1073 ff)

358. D. Movement of water molecules through the glomerulus is often greater than that accounted for by simple net diffusion. This "streaming" of molecules is termed bulk flow. Bulk flow is a characteristic of extremely permeable membranes such as those found in the renal glomerulus. (REF. 1, p. 96 ff)

359. D. The major fraction of urinary ammonia is derived from the amide nitrogen of glutamine. (REF. 1, p. 446 ff)

360. D. Polyuria associated with a high blood glucose level is an osmotic event that will affect water loss to a much greater extent than electrolyte loss. (REF. 1, p. 421)

361. E. The renal plasma threshold for the appearance of glucose in the urine will be decreased by an increase in glomerular filtration rate, a decrease in glucose T_m, decreased tubular reabsorption of glucose, and a decrease in the slope of the glucose reabsorbtion curve. (REF. 2, p. 1072 ff)

362. D. Because of the osmotic gradient from the outer cortex to the inner medulla of the kidney, the urine as it enters the collecting ducts, can only be hypotonic, or at best isotonic, but not hypertonic. (REF. 2, p. 1080 ff)

363. B. An increase in extracellular fluid osmolality (excess Na^+ and its associated anions) stimulates the osmoreceptors within the supraoptic nuclei of the hypothalamus. Impulses from the nuclei traverse through the pituitary stalk into the posterior pituitary gland promoting the release of ADH. In addition, the extracellular fluid osmolality increases and by definition means that extracellular fluid volume is less than normal. The reduction of stretch, which follows, with the atria thereby reduces nerve signals into the brain to cause an increase in ADH secretion. (REF. 1, p. 426 ff)

364. A. All are correct except D. Effective renal plasma flow describes the delivery of plasma to the peritubular capillaries. (REF. 2, p. 1061 ff)

365. B. The juxtaglomerular apparatus found in juxtamedullary nephrons which have long loops of Henle is important in body sodium homeostasis and aldosterone secretion. It is composed of the macula densa, extraglomerular mesengeal cells, and granular cells. (REF. 2, p. 1061 ff)

366. B. As soon as an individual is placed on a low sodium diet, the normal mechanisms of the body will attempt to conserve the sodium available. Hence, renin and aldosterone will be high and plasma sodium concentration will be about normal. (REF. 2, p. 1075 ff)

367. B. A subject ingesting distilled water will have a rapid decrease in plasma and urine osmolality because of dilution. In addition, the individual receiving distilled water will have a more rapid increase in urine production because the system controlling plasma osmolality works more rapidly than the system controlling plasma volume. (REF. 1, p. 390 ff)

368. D. Sodium represents by far the largest portion of tubular reabsorption. (REF. 1, p. 472 ff)

369. A. The excretion of hypertonic urine utilizing the countercurrent multiplier mechanism depends on the anatomic arrangement of the loop of Henle, the vasa recta, and collecting ducts going through the hypertonic medulla in order to allow the urine to become concentrated by equilibrating (in the presence of ADH) with the medullary interstitial fluid. (REF. 1, p. 414 ff)

370. E. Each of the factors stimulates aldosterone secretion. (REF. 2, p. 1508 ff)

371. B. Angiotensin II as well as total plasma volume are potent signals for activation of the zona glomerulosa to release aldosterone. (REF. 3, p. 512 ff)

372. B. Renal clearance is defined as the volume of plasma containing the amount of a substance that is excreted in the urine. (REF. 2, p. 1072 ff)

373. C. Renal blood flow through both kidneys accounts for one fourth of the cardiac output. (REF. 1, p. 395)

374. D. ADH is one of the primary mechanisms for the control of plasma osmolality. (REF. 2, p. 1079 ff)

375. D. When plasma osmolality and plasma sodium are low, there should be an excretion of water. If, however, the urine osmolality remains high in spite of the low plasma values, inappropriate ADH secretion should be considered. (REF. 2, p. 1079 ff)

376. B. Renal blood flow $= \dfrac{\text{Renal plasma flow}}{1 - \text{Hematocrit}}$, thus from the values, RBF $= 600/(1 - 0.40) = 1000$ mL blood/min. (REF. 1, p. 1062 ff)

377. C. According to measurements made with inulin, the GFR in man is about 125 mL/min. In the female it is about 10% less. To compare the GFRs in individuals of various sizes it is customary to normalize the values to that of an ideal person with a body surface of 1.73 cm^2. (REF. 3, p. 857)

378. D. The normal rate of urine flow is 1 mL/min with an osmolality of around 1000 mOsm/kg H_2O. If 1 L of water is ingested, the urine flow increases within 20 mm, and peaks within 1 hour. The urine osmolality is inversely related to the rate of urine flow. Since both GFR and rate of solute excretion remain the same, the change in water excretion results from differences in tubular reabsorption of water, and hence, is due to a decreased secretion of ADH. (REF. 1, p. 390)

379. A. The glomerular filtration rate can be calculated from the clearance in inulin, since it is neither excreted nor absorbed by the renal tubules.

$$\text{GFR} = \frac{\text{Inulin concentration in urine} \times \text{Urine flow}}{\text{Concentration of inulin in plasma}}$$

(REF. 2, p. 1069 ff)

380. C. Renal plasma flow can be calculated if a material is available which is not metabolized by the kidney and if the amount in the urine and loss per liter of plasma are known.

$$\text{Renal plasma flow} = \frac{\text{Amount urine/unit time}}{\text{Loss/L plasma}}$$

(REF. 2, p. 1068 ff)

381. D. Renal blood flow can be calculated from renal plasma flow and the hematocrit. (REF. 2, p. 1068 ff)

382. A. Normally, more than 99% of the filtered water is reabsorbed as it passes through the tubules, of which 65% (55 mL/min of the cleared 12 L/min) is reabsorbed in the proximal portion passively by osmosis. As a substance is reabsorbed (passively or actively) the concentration within the proximal tubule decreases, causing water to move out of the tubule by osmosis. (REF. 1, p. 404 ff)

383. C. The diluting segment of the tubules includes the ascending limb of the loop of Henle and about one half of the convoluted portion of the loop of Henle is the major site for salt conservation. As the name implies, the function of the diluting segment is to dilute the tubular fluid. But really, the cells are specifically adapted for active transport of Cl^- ions from inside the tubular lumen into the peritubular fluid. This transport of negative chloride and positive sodium ions outward creates a net $+6$ mV charge inside the tubule, which causes Na^+ ions to diffuse out from lumen to peritubular fluid. This segment is also impermeable to water and more impermeable to urea. Hence, the remaining fluid is very dilute (except urea which is high). (REF. 2, p. 1061 ff)

384. A. Normally, all of the amino acids (as well as glucose) are not directly reabsorbed by active processes in the proximal tubule. They are cotransported with sodium ions. Specifically, the electrochemical gradient for sodium entry provides the energy for glucose and amino acid transport across the brush border. (REF. 2, p. 1074)

385. A. About 99% of the filtered Na^+ is reabsorbed by all the renal tubules. Approximately two thirds of the filtered sodium is reabsorbed from the proximal tubule. The epithelial cells which line the proximal tubules have a "brush" border composed of very small microvilli. At the base of each cell are basal channels. The electrical potential within the cell is about -70 mV. Active transport of Na^+ occurs from inside the epithelial cell into the basal channels as well as into the spaces between the cells. This outward Na^+ transport reduces the Na^+ concentration inside, and because of the low concentration inside the cell, there is a Na^+ concentration gradient between the cell and the tubular lumen fluid. As a result, Na^+ will diffuse from the tubule through the brush border into the cell, where it is actively carried into the peritubular fluid of the basal channels. So both the inside negative potential and the concentration gradient cause Na^+ to diffuse from the tubular lumen into the cell. This electrochemical gradient accounts for proximal Na^+ diffusion and amounts to about 65% of the filtered load. (REF. 2, p. 1061)

386. B. The excretion of excess solutes and, hence, a concentrated urine is dependent on first creating a hyperosmolality of the medullary interstitial fluid. Normally, body fluid osmolality is 300 mOsm/L, while in the medullary tubules it approaches 1200 mOsm/L. In the thick ascending limb this active extrusion of Cl^- plus passive electrogenic absorption of Na^+ results in an increased medullary osmolality. These, along with K^+ and C^{2+}, are carried downward into the inner medulla by the blood in the vasa recta. Ions are also transplanted from the collecting duct into the medullary interstitial fluid, mainly from active transport of Na^+ and electrogenic passive absorption of Cl^- along with the Na^+. In addition, when ADH concentration is high in the blood, large amounts of urea are also reabsorbed into the medullary fluid from the collecting duct. Finally, there is the passive transport of Na^+ and Cl^- into the inner medullary interstitium from the thin segment of the loop of Henle. This passive movement results from the high urea concentration in the medullary interstitium around the collecting ducts and promotes water osmosis out of the descending limb. As a result, there is a high NaCl concentration to twice normal inside the descending thin loop. Because of the high NaCl concentration, the ions move passively out of the thin segment and into the interstitium. All of these factors cause a marked increase in the medullary interstitial fluid and are referred to as the "countercurrent" fluid flow in the loop. As can be noted then, there may be occasions when the collecting duct has an equally high fluid concentration as the loop, but it could be lower. The loop is always hyperosmotic. (REF. 2, p. 1061 ff)

387. C. The thick ascending section of the loop of Henle is the diluting segment of the tubule because of its high water impermeability. The cells of this area are also specifically adapted for active transport of Na^+ and Cl^- ions from inside the tubular lumen into the peritubular fluid. The low H_2O permeability and active reabsorption of Na^+ and Cl^- means that in this area both the osmolality and NaCl concentration is lowered to levels below those in the surrounding fluid. (REF. 2, p. 1061 ff)

388. B. An increase in blood volume will result in a decrease in the secretion of renin from the juxtaglomerular apparatus. (REF. 2, p. 1088)

389. B. Increased blood volume would decrease the presence of the angiotensin by decreasing the renin-assisted breakdown of globulin. (REF. 2, p. 1102)

390. B. The adrenal cortex is stimulated by renin. Increased blood volume decreases renin release, hence, decreases adrenal cortical secretion. (REF. 2, p. 927 ff)

391. B. The aldosterone secretion of the adrenal cortex is related to renin secretion and blood volume through a negative feedback relationship. (REF. 2, p. 927 ff)

392. B. ADH secretion will be decreased by increased blood volume. This will allow the excretion of Na^+ and water to return blood volume toward normal. (REF. 2, p. 1079 ff)

393. B. The major function of ADH is to cause a decrease in urinary excretion of water (and Na^+) in order to return blood volume towards normal. (REF. 2, p. 1080 ff)

394. A. Lower circulating levels of aldosterone will result in increased K^+ reabsorption. (REF. 1, p. 420)

395. B. Water reabsorption will be inhibited by increased blood volume due to decreased secretion of ADH and aldosterone. (REF. 2, p. 1080 ff)

396. B. The movement of fluid into the vascular system during the early stages of hemorrhagic shock causes the hematocrit to decrease. (REF. 1, p. 327 ff)

397. B. Decreased venous return will result in decreased cardiac output during shock. (REF. 1, p. 327 ff)

398. A. Peripheral resistance is increased during early stages of shock in an attempt to maintain normal blood pressure. (REF. 1, p. 327 ff)

399. A. Plasma protein concentration goes up during later stages of shock because of the release of protein from dying cells. (REF. 1, p. 327 ff)

400. B. The precapillary to postcapillary resistance ratio is decreased during irreversible shock. (REF. 1, p. 327 ff)

401. B. Mean arterial blood pressure is reduced during irreversible shock because the normal mechanisms for maintaining blood pressure are overwhelmed. (REF. 1, p. 327 ff)

402. E. Total body water makes up about 60% of the lean body weight. (REF. 2, p. 1098 ff)

403. B. Interstitial compartment water comprises approximately 15% of total lean body weight. (REF. 2, p. 1098 ff)

404. D. Intracellular water makes up approximately 40% of lean body weight. (REF. 2, p. 1098 ff)

405. A. Plasma water content comprises about 4% of lean body weight. (REF. 2, p. 1098 ff)

406. C. Extracellular fluid makes up about 20% of lean body weight. (REF. 2, p. 1098 ff)

407. A. Respiratory acidosis is caused by retention of CO_2 and may be produced by inhalation of gas mixture with high CO_2 content, voluntary breath holding, pulmonary insufficiency (as in emphysema), respiratory obstruction, respiratory center depression, or by paralysis of the respiratory muscles. The change in the blood is an increase in arterial hydrogen ion concentration and CO_2 tension. (REF. 2, p. 1096)

408. A. Because of the increased blood pCO_2 there will be an increase in HCO_3^- content during respiratory acidosis. The resulting renal excretion of chloride will eventually lead to an increase in buffer base. (REF. 2, p. 1094 ff)

409. B. Respiratory alkalosis occurs when alveolar pCO_2 is lower than normal and results from an increase in alveolar ventilation. The effect on the blood is to decrease arterial hydrogen ion concentration and CO_2 tension. Causes of this condition include voluntary hyperventilation, anoxemia, hysteria, fever, dyspnea due to congestive heart failure, and lesions involving the brain stem. (REF. 2, p. 1094 ff)

410. B. Following the lowered pCO_2, during respiratory alkalosis there will be a decrease in HCO_3^- content of the blood. The concomitant increase in plasma Cl^- concentration leads eventually to increased renal excretion of base, which lowers the buffer base content. (REF. 2, p. 1094 ff)

411. B. Metabolic acidosis may be caused by an excess of fixed acid associated with metabolism, ingestion of acidifying salts, or by any condition in which more anions than cations are present in the circulation: the characteristic finding in metabolic acidosis is diminution in total base, decrease in buffer base, and an increase in arterial hydrogen ion content while CO_2 tension is reduced. (REF. 2, p. 1094 ff)

412. B. As a result of the lower blood pCO_2 during metabolic acidosis, the HCO_3^- concentration will decrease. The condition is usually observed after massive diarrhea, particularly in infants. (REF. 2, p. 1094 ff)

413. A. Metabolic alkalosis may occur either by a deficiency of fixed base (as following persistent vomiting) or by continuous loss of gastric juice. While the CO_2 content is raised following decreased ventilation, the total base and buffer base are decreased with an increase in plasma pH. The increase in blood pCO_2 follows the attempt by the lungs to restore the blood pH to normal. (REF. 2, p. 1094 ff)

414. A. Following the increase in blood pCO_2 there will be an increase in plasma HCO_3^- during metabolic alkalosis. The arterial pH and HCO_3^- increase together, which enables metabolic alkalosis to be distinguished from respiratory alkalosis. (REF. 2, p. 1094 ff)

6 Cardiovascular Physiology

DIRECTIONS (Questions 415–453): For each of the questions or incomplete statements below, **one** or **more** of the answers or completions given is correct. Select

    A if only *1, 2, and 3* are correct
    B if only *1 and 3* are correct
    C if only *2 and 4* are correct
    D if only *4* is correct
    E if all are correct

415. A ventricular function curve refers to the relationship between
1. stroke work and fiber length
2. stroke volume and heart rate
3. stroke work and filling pressure
4. cardiac output and stroke work

416. In heart failure there is
1. inadequate cardiac output
2. increased tissue blood flow
3. elevated venous pressure
4. adequate fluid tissue exchange

417. The peak velocity of blood flow in the ascending aorta is
     1. about 10 cm/s
     2. less than that in the carotid artery
     3. much slower than in capillaries
     4. inversely proportional to the cross-sectional area

418. Increased pressure within the carotid sinus causes
     1. a decrease in sympathetic tone
     2. a fall in venous pressure
     3. reflex bradycardia
     4. reflex hyperpnea

419. Within the jugular pulse, the C wave
     1. occurs just after the peak of auricular systole and just prior to the dicrotic notch
     2. occurs during ventricular systole
     3. occurs at the onset of expiration
     4. is the result of bulging of the mitral and tricuspids into the aorta

420. Hemorrhagic shock will result in a decrease in all of the following EXCEPT
     1. cardiac work
     2. cardiac output
     3. coronary flow
     4. plasma pCO_2

421. Anemia due to maturation failure of the red blood cell (RBC) is regarded as
     1. lack of intrinsic factor
     2. reduced cobalamin absorption
     3. atrophy of gastric mucosa
     4. pernicious anemia

422. In hemorrhagic shock one would expect a decrease in
     1. cardiac work
     2. cardiac output
     3. coronary flow
     4. oxygen usage

Directions Summarized				
A	B	C	D	E
1,2,3	1,3	2,4	4	All are
only	only	only	only	correct

423. Vasomotor reflexes may be mediated through special areas of the
    1. cerebrum
    2. hypothalamus
    3. medulla
    4. spinal cord

424. If one of the cardiac ventricles becomes hypertrophied, it will depolarize
    1. before the other ventricle
    2. more rapidly than the other ventricle
    3. with electrical properties identical to those of the other ventricles
    4. and shift the electrical axis to the involved side

425. Changes in the mean electrical axis of the ventricles may be caused by
    1. changes in body position
    2. hypertrophy of one ventricle
    3. muscular necrosis
    4. bundle branch block

426. If the end-diastolic ventricular volumes are increased (within physiologic limits)
    1. the force of cardiac contraction is increased
    2. the stroke volume is increased
    3. cardiac output is increased
    4. Starling's law is applicable

427. In Figure 16, which of the following hemodynamic situation(s) exist(s)?
   1. The blood flow profile is streamline (laminar) in nature
   2. Lamina A is moving faster than lamina B, C, or D, with the slowest velocity occurring in D
   3. The greatest pressure in this vessel exists at the vessel walls
   4. The greatest number of RBCs will flow in lamina A

428. Turbulent blood flow within the aorta occurs
   1. at high flow velocities when the streamline flow breaks down
   2. when the fluid particles move in irregular and varying paths, forming eddies
   3. if and whenever the viscosity and density of the blood are altered
   4. following any change in blood vessel diameter

429. The first heart sounds consist of vibrations
   1. of the arteriovenous (AV) valves during and after their closure
   2. set up by eddy currents in the blood ejected through the aortic valve
   3. emanating from the heart muscle fibers themselves
   4. resulting from the closure of the semilunar valves

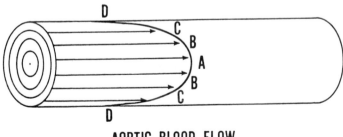

AORTIC BLOOD FLOW

Figure 16. Blood flow in a large artery such as the aorta.

Directions Summarized				
A	**B**	**C**	**D**	**E**
1,2,3	1,3	2,4	4	All are
only	only	only	only	correct

E **430.** Cardiac reserve refers to the increase above the ordinary level that can be attained in
1. heart rate
2. cardiac output
3. cardiac work
4. stroke volume

D **431.** In general, in the response to moderate exercise, in the untrained individual
1. stroke volume decreases with a constant heart rate
2. heart rate increases with an increased stroke volume
3. heart rate increases with a decrease in stroke volume
4. heart rate increases with a constant stroke volume

E **432.** Extrinsic control of myocardial function may be exerted by the
1. autonomic nervous system
2. catecholamine content of the blood
3. oxygen content of the myocardial blood
4. carbon dioxide content of the myocardial blood

B **433.** An increase in capillary permeability, as might occur with the introduction of bacterial toxins into the circulation, would produce which of the following alterations?
1. Reduction of the plasma volume
2. Reduction of the hematocrit
3. Formation of edema
4. Reduction of flow of the lymph

434. Under normal physiologic conditions, the capillaries 5%
   1. contain more than 25% of the total blood volume
   2. have a very rapid blood flow
   3. have a higher blood flow and pressure than the arterioles
   (4.) contain a relatively small amount of the total blood volume (about 300 mL)

435. In Figure 17,
   1. curve B expresses a decreased capability of the heart to serve as a pump
   2. curve A expresses a contractile capability of the heart that enables it to maximally respond to changes in venous return
   3. curve C is an expression of cardiac capability that has been "improved" and is returning to "normal"
   4. curves A, B, and C are expressions of a relationship between fiber length and muscle tension

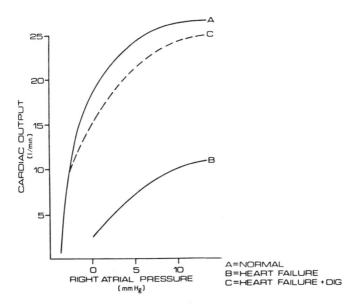

Figure 17. Three ventricular functional curves (A, B, C), representing the ability of the heart to function under various conditions.

Directions Summarized

A	B	C	D	E
1,2,3	1,3	2,4	4	All are
only	only	only	only	correct

436. Under resting conditions the cardiac output of humans would be closest to   *old people too*
   1. 2.0 L/min
   2. 4.0 L/min
   3. 7.5 L/min
   4. 5.0 L/min

437. In the normal hemodynamic situation, which of the following relationships operate (see Figure 18)?
   1. P_T, exerted on the wall, is the difference between P_I and P_E
   2. The relationship between pressure and wall tension in the vessel is such that the tension is determined by P/r
   3. The relationship between pressure and wall tension is largely determined by vessel radius and may be expressed as $T = Pr$
   4. P_I exerted on the wall is independent of P_E

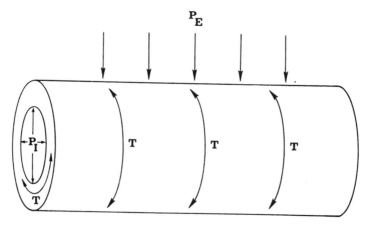

Figure 18. The forces (arrows) acting on a blood vessel: (P_I) intravascular pressure; (P_E) extravascular pressure; (T) circumferential tension in the vessel wall.

438. The parallel arrangement of most of the tissue circulations is such that blood flow to a tissue may *Independent of CO*
 1. increase as cardiac output increases
 2. decrease in spite of an increased cardiac output
 3. be maintained in the face of a reduced cardiac output
 4. be maintained in the face of a reduced stroke volume

439. The transmembrane potential in cardiac tissue
 1. depends on sodium ion concentration
 2. depends on the balance between chemical and electrostatic forces of a resting cardiac cell
 3. depends on the cell membrane permeability to sodium, potassium and calcium ions
 4. is independent of sodium ion concentration

440. The excitability of a cell depends on
 1. the amplitude of the action potential
 2. resting membrane potential
 3. a sudden decrease in g_{Na}
 4. the extracellular concentration of sodium ions

441. In a normal cardiac cell at rest
 1. V_m is approximately -90 mV for myocardial fibers in the atria and ventricles
 2. V_m is approximately -90 mV for Purkinje fibers
 3. the intracellular to extracellular ratio of potassium ions is approximately $30:1$
 4. g_K is about 100 times greater than g_{Na}

442. Regarding the cardiac cell
 1. in a resting cell, chemical and electrostatic forces are poised to draw potassium into the cell
 2. in a resting cell, chemical and electrostatic forces are poised to draw sodium into the cell
 3. gating does not occur until the threshold value of -65 mV is reached
 4. gating is a voltage-dependent phenomenon

		Directions Summarized		
A	**B**	**C**	**D**	**E**
1,2,3	1,3	2,4	4	All are
only	only	only	only	correct

443. Which of the following is true regarding conduction in cardiac tissue?
1. Current tends to flow from high potential to low potential
2. Is similar to nerve conduction in that the fluids on either side of the cell membrane are good conductors of electricity
3. True current is generated by movement of anions in one direction with movement of cations in the opposite direction with respect to either side of the cell membrane
4. Propagation varies ~~inversely~~ with the action potential amplitude

444. In conduction of the fast response
1. opening of the m gates is independent of the rate of sodium flux ⟋ ₍want Na⁺₎
2. opening of the m gates varies directly with changes in V_m
3. action potential amplitude varies directly with opening of the m gates, but is independent of the rate of change of potential
4. the amplitude of the action potential and the magnitude of the current at any given point along a fiber is proportional to the potential difference across the membrane

445. In conduction of the slow response
1. the threshold is approximately −67 mV
2. conduction may proceed in one direction but is frequently blocked in the other direction, a condition known as undirectional block which forms the basis for many arrhythmias
3. as V_m becomes less negative, h gates tend to open
4. refractory period for h gates is much longer than for m gates

446. Third degree atrio-ventricular block is a condition in which
1. there is a complete block of conduction between the atria and the ventricles
2. each QRS is preceded by a P wave
3. atrial complexes usually occur at a rate different from that of ventricular complexes
4. the P-R interval is constant at .12

447. Ectopic foci or ectopic pacemakers
1. will initiate beats when conduction pathways between ectopic foci and regions of higher degree rhythmicity are blocked
2. include the sinoatrial (SA) node
3. can be found in all four chambers of the heart
4. are independent of those intrinsic features associated with automaticity and rhythmicity

448. The sinoatrial node
1. is considered the natural pacemaker region of the heart
2. has a lower (less negative) resting potential
3. action potential is unaffected by tetrodotoxin
4. repolarization (phase 3) is slower and more gradual

449. The rate or frequency of discharge of pacemaker cells may vary due to changes
1. in rate of depolarization during phase 4
2. in the threshold potential
3. in the resting potential
4. in the depolarization of phase 2

		Directions Summarized		
A	**B**	**C**	**D**	**E**
1,2,3	1,3	2,4	4	All are
only	only	only	only	correct

450. During phase 2 of the action potential for a Purkinje fiber
1. there is an influx of sodium ions through the fast channels
2. there is an influx of mainly calcium ions
3. there is an increase in g_K
4. there is an increase in the g_{Ca}

451. Regarding reentry
1. cardiac impulses reexcite regions previously passed
2. it may be ordered or random
3. it requires that at some point there is unidirectional blockage
4. fibrillation is a typical example

452. Which is not a determination based on electrocardiography?
1. Mechanical performance of the heart
2. Cardiac output
3. Systolic and diastolic information
4. Effects due to changes in fluid electrolytes

453. Characteristics indicative of third-degree AV block include:
1. independent atrial and ventricular rhythm
2. complete heart block
3. no fixed relationship between QRS and P waves
4. usually associated with syncope

DIRECTIONS (Questions 454–512): Each of the questions or incomplete statements below is followed by five suggested answers or completions. Select the **one** that is **best** in each case.

454. The term hematocrit means the
    A. percentage of the blood that is red blood cells
    B. percentage of blood that is plasma
    C. ratio of the blood volume to the extracellular space
    D. percentage of new blood formed every 120 days
    E. Average specific gravity of formed elements

455. A number of different chemical substances cause leukocytes to move either toward or away from the source of the chemical. This phenomenon is known as
    A. hydrotropism
    B. chemotropism
    C. diapedesis
    D. inflammatory response
    E. chemotaxis

456. The pulse pressure is highest in the
    A. aorta
    B. femoral artery
    C. dorsalis pedis
    D. arterioles
    E. capillaries

457. Within physiologic limits, an increase in right atrial pressure
    A. decreases systemic arterial pressure
    B. increases cardiac output
    C. decreases intrathoracic pressure
    D. decreases heart rate
    E. decreases renal blood flow

458. Essential hypertension is generally associated with an early increase in
   A. cardiac work
   B. cardiac output
   C. coronary flow
   D. oxygen usage
   E. cardiac efficiency

459. In congestive heart failure edema results from
   A. an elevation of cardiac output
   B. an elevation of peripheral resistance
   C. a reduction of plasma oncotic pressure
   D. an elevation of capillary hydrostatic pressure
   E. an increase in tissue oncotic pressure

460. In left bundle branch block of cardiac excitation there will be heart sound changes such that the second sound is
   A. louder than normal
   B. widely split during inspiration
   C. widely split during expiration
   D. paradoxically or reversibly split during expiration because of the delay in closure of the aortic valve
   E. very hard to hear

461. If the heart rate of a resting adult is 160 beats/min, the arterial pressure is 130/70, and the central venous pressure is high (8 mm Hg), one might reasonably suspect
   A. essential hypertension
   B. internal bleeding
   C. a large arteriovenous shunt
   D. massive sympathetic activity
   E. massive parasympathetic activity

462. An increase in stroke volume alone, without a change in peripheral resistance, heart rate, or arterial capacitance, will result in
   A. an increase in pulse pressure and an increase in mean pressure
   B. an increase in pulse pressure and a decrease in mean pressure
   C. a decrease in pulse pressure and an increase in mean pressure
   D. an increase in both systolic and diastolic pressures, no change in mean pressure
   E. no change in either pulse or mean pressure

463. Increasing radius from 1 to 2 cm should cause resistance to flow to
   A. double
   B. half
   C. increase 16 times
   D. decrease 16 times
   E. decrease eight times

464. The order of valve opening and closing during a single cardiac cycle is as follows (M = mitral, T = tricuspid, A = aortic, P = pulmonic):
   A. M closes, T closes, A opens, P opens, P closes, A closes, M opens, T opens
   B. M closes, T closes, A opens, P opens, A closes, P closes, T opens, M opens
   C. T closes, M closes, A opens, P opens, A closes, P closes, T opens, M opens
   D. T closes, M closes, P opens, A opens, A closes, P closes, M opens, T opens
   E. P closes, A closes, M opens, T opens, M closes, T closes, P opens, A opens

465. Tissue blood flow will increase as a result of
   A. increased pO_2
   B. reduced plasma (K^+)
   C. increased pCO_2
   D. decreased plasma osmolality
   E. decreased blood pressure

466. The amount of blood present in the liver is determined predominantly by the
    A. portal pressure
    B. hepatic venous pressure
    C. arterial pressure
    D. hepatic tissue protein
    E. serum protein

467. Normally, if the venous pressure increases by 8 mm Hg, the capillary pressure will
    A. not change
    B. decrease by 8 mm Hg
    C. increase by 8 mm Hg
    D. decrease by 6 mm Hg
    E. increase markedly

468. If two vessels are connected in parallel, their total resistance to blood flow is
    A. more than if they were connected in series
    B. the sum of their individual resistances
    C. the average of their individual resistances
    D. the same as the resistance of the smaller of the two
    E. less than the resistance of either vessel alone

469. The average velocity of blood flow in the femoral artery is
    A. about 10 cm/s
    B. less than that in the carotid artery
    C. independent of the cardiac output
    D. inversely proportional to the cross-sectional area
    E. above that at which turbulence develops

470. With constant flow through a tube system
    A. reducing the diameter of the tube will reduce lateral pressure
    B. increasing tube diameter will reduce lateral pressure
    C. reducing tube diameter will reduce flow velocity
    D. increasing tube diameter will increase flow velocity
    E. velocity is not affected by tube size

471. An increase in ventricular stroke work is obtained by administering a new pharmacologic drug claimed to be a positive inotrope. If the heart rate is not changed significantly, and if the effective (end-diastolic) filling pressure is not increased, the effect of this drug also might represent
   A. homeometric autoregulation associated with an increase in mean aortic pressure, following peripheral vasoconstriction
   B. increased parasympathetic (vagal) activity
   C. heterometric (Frank-Starling) autoregulation
   D. heterometric autoregulation due to increased vagal activity
   E. none of the above

472. If arterial capacitance remains constant, increasing heart rate should cause pulse pressure to
   A. increase
   B. decrease
   C. remain unchanged
   D. decrease then increase
   E. increase then decrease

473. In hemodynamics peripheral resistance is
   A. the resistance to exercise so as to aid circulation
   B. the resistance to blood flow through the systemic circulatory system
   C. peripheral contraction of skeletal muscle which stops blood flow
   D. the dilation of peripheral vessels which causes pooling of blood
   E. resistance to the flow of blood through the lungs

474. To determine cardiac output by the dye-dilution technique it is necessary to measure dye concentration
   A. from a sample from the aorta
   B. continuously from arterial blood
   C. from a sample from the right atrium
   D. from a sample from a peripheral artery
   E. from a sample from the saphenous vein

475. Stimulation of the midline vasodilator area of the vasomotor center
  - A. inhibits the neighboring vasoconstrictor area
  - B. sends fibers to the heart with the vagus nerve
  - C. inhibits transmission through sympathetic ganglia
  - D. inhibits impulses through the vagus nerve
  - E. affects the aorta but not the other large arteries

476. The mean stroke volume is determined by assessing the
  - A. cardiac output in liters per minute divided by the heart rate in beats per minute
  - B. cardiac index divided by the O_2 extraction ratio
  - C. cardiac efficiency divided by the cardiac index
  - D. mean arterial pressure divided by the mean stroke volume
  - E. cardiac output (L/min) divided by the cardiac index

477. The direct Fick method for determination of cardiac output in man requires measurement of
  - A. O_2 consumption per unit time only
  - B. O_2 content of arterial blood only
  - C. O_2 content of blood taken from the right ventricle only
  - D. the arteriovenous oxygen difference and content of the arterial blood
  - E. O_2 content of right ventricular blood as well as the arteriovenous O_2 difference

478. Decreased pressure within the carotid sinus causes
  - A. an increase in heart rate
  - B. a fall in venous pressure
  - C. reflex bradycardia
  - D. reflex hyperpnea
  - E. reflex increase in venous pressure

479. Autoregulation of blood flow involves the maintenance of a relatively constant
  - A. blood pressure during variations of blood flow
  - B. resistance in the face of changing blood pressure
  - C. blood flow over a range of blood pressure
  - D. blood flow while resistance is changing
  - E. heart rate in the face of changing blood pressure

480. Blood flow is regulated largely by local metabolic effects in
    A. brain
    B. muscle
    C. skin
    D. bone
    E. kidney

481. The incisura or dicrotic notch in the aortic pressure curve is
    A. of no diagnostic value
    B. absent in arteriosclerosis
    C. indicative of cardiovascular pathology
    D. coincident with the second heart sound
    E. magnified by aortic regurgitation

482. During moderate exercise the heart rate increases with a(n)
    A. decreased stroke volume
    B. constant stroke volume
    C. increased stroke volume
    D. initial stroke volume increase, then decreased stroke volume
    E. moderate increase in cardiac efficiency

483. The portion of the circulatory system with the largest total cross-sectional area is the
    A. large arteries
    B. arterioles
    C. capillaries
    D. small veins
    E. large veins

484. If 100,000 cpm of radioiodinated albumin and 200,000 cpm of radiolabeled red blood cells were injected intravenously and allowed to completely mix in the vascular system without loss, and if the equilibrium concentrations were determined to be 100 cpm/mL of plasma for the plasma indicator and 400 cpm/mL of red cells for the red cell indicator, the plasma volume would be (in mL)
  A. 1000
  B. 250
  C. 2000
  D. 500
  E. 100

485. Reducing carotid sinus transmural pressure will result in
  A. an increased peripheral resistance
  B. a reduction in cardiac contractility
  C. a reduction in cardiac output
  D. a decrease in peripheral resistance
  E. no change in peripheral resistance

486. A rise in the blood pressure within the carotid sinus results in a(n)
  A. reduction in vagal discharge to the heart
  B. increase in sympathetic discharge to the heart
  C. increased venomotor activity
  D. increased parasympathetic activity
  E. increase in ADH levels

487. The cardiac index, by definition, is the ratio of
  A. cardiac output and body surface area
  B. cardiac output and body weight
  C. cardiac output and work of the heart
  D. stroke volume and surface area of the body
  E. body surface area and heart rate

488. The pulse pressure is lowest in the
  A. aorta
  B. femoral artery
  C. dorsalis pedis
  D. arterioles
  E. capillaries

489. A sudden increase in right atrial volume will cause an immediate
   A. increase in systemic arterial pressure
   B. increase in the pressure within the thorax
   C. decrease in cardiac output
   D. increase in the heart rate
   E. increase in cardiac output

490. In a progressive AV block one would expect
   A. progressive prolongation of the interval between atrial and ventricular contractions
   B. progressive shortening of the interval between atrial and ventricular contractions
   C. progressive weakening of atrial contractions
   D. progressive weakening of ventricular contractions
   E. progressive strengthening of ventricular contractions

491. Contraction of the splenic capsule in man
   A. increases the circulating blood volume
   B. increases the hematocrit
   C. occurs with exercise
   D. is not known to occur because of its lack of musculature
   E. results from sympathetic stimulation

492. The venous pulse in the greater veins is
   A. a dampened arterial pulse
   B. a result of pressure changes in the heart and neighboring arteries
   C. necessary for proper cardiac filling
   D. diminished in cardiac failure
   E. abolished at heart rates of over 150 beats/min

493. Bleeding time
   A. measures the rate of bleeding from a large puncture
   B. is normally about 2 to 3 minutes
   C. is significantly longer in hemophilia
   D. is independent of platelet concentration
   E. is normally about 6 to 8 minutes

494. Latent pacemaker activity in the heart resides in the
- **A.** Left ventricular muscle
- **B.** bundle of His
- **C.** N layer of the AV node
- **D.** the Purkinje network
- **E.** left atrial muscle

495. It is impossible to tetanize a heart because
- **A.** there is a long mechanical refractory period
- **B.** the electrical refractory period and the mechanical contractile response are of almost equivalent duration
- **C.** the mechanical contractile event is usually shorter than the duration of the electrical depolarization
- **D.** the Ca^{2+} transport mechanism in heart muscle is responsible for the prolonged refractoriness
- **E.** heart muscles do not contain Ca^{2+}

496. During the reduced ejection phase of the left ventricle, which one of the following is true?
- **A.** Left atrial pressure is falling
- **B.** Aortic flow velocity is rapidly decrementing
- **C.** Aortic pressure is falling below left ventricular pressure
- **D.** Left ventricular pressure is constant
- **E.** The tricuspid valves are closed

497. Both cyanosis and systemic arterial blood hypoxemia may occur in the presence of
- **A.** potassium cyanide (KCN) poisoning
- **B.** a low cardiac output
- **C.** regions of abnormally low ventilation/perfusion (V/Q) ratio
- **D.** some regions of the lung being relatively more perfused than others
- **E.** heart rates between 40 and 50 beats/min

498. Pressure in the main pulmonary artery
   A. will approximately double if cardiac output doubles
   B. will approximately double if one lung is removed
   C. is lowered by a local vasodilator effect of alveolar hypoxia
   D. is always high enough to perfuse the uppermost parts of a human lung
   E. is not inversely proportional to cardiac output

499. Cardiac work is most nearly equal to
   A. area of the pressure-volume diagram
   B. tension-time index
   C. kinetic energy imparted to movement of blood
   D. systolic blood pressure
   E. heart rate

500. The sympathetic adrenergic stimulation results in
   A. vasoconstriction in all tissues
   B. increased coronary blood flow
   C. increased muscle blood flow
   D. increased skin blood flow
   E. vasodilation in all tissues

501. According to the myogenic theory of autoregulation of blood flow, increasing arteriolar blood pressure leads to
   A. an elevation of tissue blood flow
   B. a decreased vascular resistance
   C. an increased vascular resistance
   D. decreased vascular tone
   E. decreased blood flow

502. The most important factor in improving the ability of the heart to increase blood flow to peripheral tissue in exercise is
   A. increased oxygen extraction from the blood by the heart
   B. increased myocardial efficiency independent of O_2 needs of the organ
   C. increased coronary blood flow
   D. increased venomotor tone
   E. decreased circulating catecholamines

503. According to the Hagen-Poiseuille relationship, doubling vessel length should cause blood flow to
    A. double
    B. halve
    C. increase 16 times
    D. decrease
    E. not change, since there is no relationship between length and blood flow

504. The length-tension relationship in cardiac muscle allows the heart to make automatic adjustments in its output. This is accomplished by
    A. a decrease in the potential force of contraction as the resting (diastolic) fiber length is increased (over a normal physiologic range)
    B. an increase in the amount and rate of shortening following an increase in resting (diastolic) fiber length (over a normal physiologic range)
    C. a reduction in the contractility of the heart muscle following an increased diastolic filling
    D. a change in the amount of actin and myosin per fiber
    E. alterations in the Ca^{2+} uptake by the sarcoplasmic reticulum

505. The calcified, rigid aorta often seen in elderly people requires an increased cardiac performance for an adequate cardiac output because it
    A. has a smaller than normal diameter and so increases the resistance to outflow from the heart
    B. accommodates less blood than normal between 90 and 150 mm Hg pressure and so requires a higher than normal ejection pressure
    C. gives less elastic recoil during diastole and so reduces the load and thus the contribution of homeometric autoregulation
    D. tends to prevent movement of the base of the heart and so reduces effective venous return
    E. decreases the amount of calcium available to the cardiac muscle

506. An adult male is seen in the clinic with the following findings:

Cardiac output,	5 L/min
Systolic pressure,	200 mm Hg
Diastolic pressure,	120 mm Hg
Heart rate,	80 bpm
Plasma catecholamines,	normal
Blood volume,	5 L
Total peripheral resistance,	increased
Plasma angiotensin,	normal

From the above data, it should be concluded that the patient is
A. hypertensive due to excessive cardiac activity
B. suffering from essential hypertension
C. in hypotension of renal origin
D. hypertensive as a result of excessive adrenomedullary secretion
E. suffering from hypertension of neural origin

507. Which of the following factors would be expected to contribute to the peripheral edema commonly observed in congestive heart failure?
A. Decreased arterial pressure, increased serum sodium, increased tissue hydrostatic pressure
B. Decreased arterial pressure, decreased venous pressure, increased hematocrit
C. Increased venous pressure, decreased serum proteins, increased plasma volume
D. Decreased arterial pH, increased skeletal muscle activity, increased sympathetic tone
E. Decreased blood volume, decreased aldosterone secretion, decreased venous pressure

508. During stress or exercise, which of the following cardiovascular system reserves have the largest potential for increasing oxygen supply to the tissues?
- **A.** Increased blood arterial oxygen content
- **B.** Increased blood flow resulting from increased stroke volume
- **C.** Increased extraction of oxygen from the blood
- **D.** Increased arterial blood pressure
- **E.** Increased venous blood pressure

509. A patient comes to you with a history and symptoms of a left myocardial infarction. The patient has an arterial blood pressure of 70/40 with a cardiac output that is one third of normal. The systemic central venous pressure would be expected to be
- **A.** increased, especially if the right ventricle is responding normally to the resultant massive sympathetic outflow
- **B.** increased, especially if the right ventricular response to the increased pulmonary vascular pressure is inadequate
- **C.** not changed, because changes in left ventricular function are not reflected to the right ventricle
- **D.** decreased, because the venous return is decreased
- **E.** decreased, because the heart rate is increased

510. If all other factors remain constant, the volume of blood flow in the aorta is
- **A.** greater than the pulse wave velocity
- **B.** approximately equal to the pulse wave velocity
- **C.** inversely proportional to aortic diameter
- **D.** directly proportional to the square of the aortic radius
- **E.** independent of the left ventricular minute volume

511. An important function of the cardiac Purkinje system is its ability to
- **A.** slow the conduction of impulses
- **B.** speed the conduction of impulses
- **C.** amplify impulses
- **D.** delay impulses
- **E.** select impulses

512. A patient is admitted to the hospital following a severe automobile accident. Upon examination you observe and obtain the following information: lower extremities are edematous, heart rate is increased, central venous pressure is decreased, total peripheral resistance is increased, venous hematocrit is increased, cardiac output is decreased. A possible cause of these findings would be

    A. anemia due to internal hemorrhage
    B. marked hysteria
    C. acute renal failure
    D. acute inferior vena caval obstruction
    E. polycythemia due to internal bleeding

DIRECTIONS (Questions 513–515): This section consists of a situation, followed by a series of questions. Study the situation and select the **one** best answer to each question following it.

Questions 513–515: The following data were obtained from two patients:

	Patient A	Patient B
Cardiac output (CO)	—	—
Mean blood pressure	—	—
Heart rate	100	150
O_2 consumption (mL/ min)	250	—
Venous O_2 (mL/100 mL)	14.5	—
Arterial O_2 (mL/100 mL)	19.0	—
Stroke volume (mL/beat)	—	70

513. The CO in patient A is (in L/min)
    a. 12.0 L/min
    B. 7.5 L/min
    C. 3.0 L/min
    D. 5.5 L/min
    E. 20.0 L/min

514. The stroke volume in patient A is (in mL/beat)
    A. 80 mL/beat
    B. 70 mL/beat
    C. 100 mL/beat
    D. 55 mL/beat
    E. 150 mL/beat

515. The CO in patient B is (in L/min)
    A. 10.5 L/min
    B. 8.0 L/min
    C. 7.0 L/min
    D. 5.5 L/min
    E. 3.0 L/min

DIRECTIONS (Questions 516–547): Each group of questions below consists of a set of lettered components, followed by a list of numbered words or phrases. For **each** numbered word or phrase, select the **one** lettered component that is most closely associated with it. Each lettered component may be selected once, more than once, or not at all.

Questions 516 and 517 (Figure 19):
    A. Curve A
    B. Curve B

516. The cardiac output of the right ventricle is represented by

517. The cardiac output of the left ventricle is represented by

Questions 518–520 (Figure 20):
    A. Site A
    B. Site B
    C. Site C
    D. Site D
    E. Site E

518. Major vascular resistance is

519. Major fluid exchange is

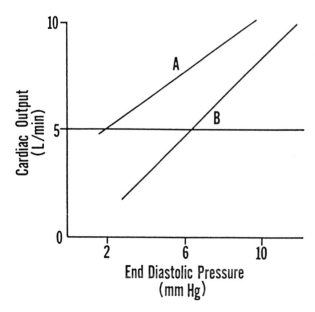

Figure 19. Relationship between cardiac output and end-diastolic pressure (mm Hg).

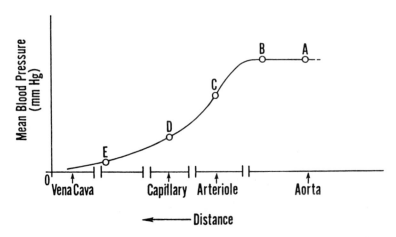

Figure 20. Changes in mean systemic blood pressure as the blood goes from the aorta through the various vessel types.

520. Lowest erythrocyte linear velocity is

Questions 521–526:
- **A.** First event
- **B.** Second event
- **C.** Third event
- **D.** Fourth event
- **E.** Fifth event
- **F.** Sixth event

Starting from the P wave, indicate the order in which the following events occur:

521. AV valve opens

522. Aortic valve closes

523. Q wave

524. T wave

525. AV valve closes

526. Aortic valve opens

Questions 527–531:
- **A.** First heart sound
- **B.** Second heart sound
- **C.** P wave
- **D.** Q wave
- **E.** T wave

527. Period of isometric contraction

528. Immediately precedes isovolumic contraction

529. Occurs during ventricular ejection phase

530. Immediately precedes atrial contraction

531. Period of isometric relaxation

Questions 532-537:

    A. Increases from normal average value

    B. Decreases from the normal average value

    C. Does not change from the normal average value

532. Mean arterial pressure in a patient with mitral stenosis

533. Left atrial pressure in a patient with aortic regurgitation

534. Cardiac output in a mitral stenotic patient

535. Left ventricular volume in a patient with mitral stenosis

536. Left atrial volume in a patient with mitral stenosis

537. Total blood volume in a patient with aortic regurgitation

Questions 538-547 (Figure 21):

    A. P wave

    B. U wave

    C. T wave

    D. QRS complex

    E. PR interval

    F. ST segment

    G. QT interval

    H. Q wave

    I. S wave

    J. R wave

538. Deflection #1

539. Deflection #2

540. Deflection #3

541. Deflection #4

542. Deflection #5

543. Deflection #6

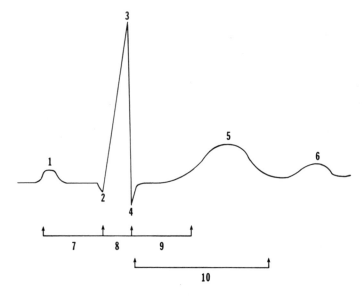

Figure 21.

544. Period #7

545. Period #8

546. Period #9

547. Period #10

Explanatory Answers

415. B. by definition, a ventricular function curve is the relationship between stroke work and filling pressure in the intact heart or fiber length from isolated tissue. (REF. 2, p. 972 ff)

416. B. In heart failure the heart cannot pump a sufficient amount of blood to meet the body's demand. The blood accumulates in the venous system, especially during exercise. (REF. 2, p. 830 ff)

417. D. The peak velocity of blood flow in the ascending aorta is about 60 cm/s and twice as fast as the descending. This is related in an inverse fashion to the cross-sectional area of the vessel. Capillary flow is the slowest and favors fluid exchange. (REF. 2, pp. 776–778, 972 ff)

418. A. The reaction to increased pressure in the carotid sinus will be all those reactions that tend to negate this pressure increase. Bradycardia is the initial reflex response due to enhanced vagal activity and reduced sympathetic activity. (REF. 3, p. 272)

419. C. The C wave occurs when the ventricles begin to contract. A large amount of blood accumulates in the atria, increasing ventricular pressure and causing bulging of the arterioventricular valves toward the atria. (REF. 1, pp. 154, 228)

420. D. As blood volume decreases all parameters decrease except pCO_2 in plasma, which accumulates from the reduction in blood flow. (REF. 2, p. 986 ff)

421. E. Pernicious anemia is due to a failure in producing red blood cells. It occurs when lack of intrinsic factor impairs cobalamin absorption leading to arrested RBC maturation. The oxyntic cells in the gastric mucosa secrete the intrinsic factor. (REF. 2, pp. 1456–1457)

422. E. During hemorrhagic shock, blood volume goes down, resulting in a decrease in all of the parameters listed. (REF. 1, p. 327 ff)

423. E. Many areas of the central nervous system are involved in the control of the cardiovascular system. (REF. 1, p. 238 ff)

424. D. When one ventricle greatly hypertrophies, the axis of the heart shifts toward the hypertrophied ventricle because far greater quantities of muscle exist on the hypertrophied side, which allows excess generation of electrical currents and more time required for the depolarization wave to travel through the hypertrophied ventricle. (REF. 1, p. 188 ff)

425. E. The mean electrical axis of the ventricles averages approximately 0.59°. A number of conditions, such as change in the position of the heart, hypertrophy of one ventricle, bundle branch block, or muscular necrosis can cause axis deviation beyond normal limits. (REF. 1, p. 188 ff)

426. E. If cardiac muscle is stretched, it develops greater contractile tension upon excitation. Increased vigor of contraction causes more external work to be done by the heart, and all the parameters listed are operative. (REF. 2, p. 824 ff)

427. E. In this diagram the flow is streamline in nature with the fluid in lamina A having the greater velocity. The velocity of flow in lamina B is greater than in C, which is faster than in D. In fact, the fluid layer in D, which is in contact with the vessel wall, moves barely or does not move at all. As a consequence of this streamline, or laminar flow, there occurs axial streaming in which the fluid's formed elements—the red blood cells—flow in the fastest layers, which will be in the center. This axial streaming is due to a velocity gradient and, according to Bernoulli's principle, the pressure exerted by the fluid against the vessel wall is related to its velocity of flow. Thus, the faster the velocity of flow, the lower the pressure exerted against the wall (the lateral pressure), and it is this pressure gradient which forces the RBCs toward the center of the vessel. (REF. 2, p. 864 ff)

428. E. All the factors listed play an important role in the type of blood flow. When laminar flow breaks down into a turbulent flow, as occurs at high velocities, the Reynolds number increases (as predicted by the Poiseuille equation). In turbulent flow the fluid particles move in constantly varying paths, and irregularly, thus forming eddy currents at some times and giving the appearance of random motion at other times. The transition from laminar to turbulent flow is greatly dependent on the density and viscosity of the fluid as well as the tube radius and the average velocity. The critical point at which turbulence develops may be determined quantitatively and referred to as the Reynolds number; i.e.,

$$\text{Reynolds number} = \frac{(\bar{v}Dp)}{n},$$

where

$\bar{v}$ = average velocity (cm/s)
D = diameter of tube (cm)
p = blood density (usually is about 1.05 g/cm^2)
n = blood viscosity (dyne s/cm^2)

Usually turbulence develops at a Reynolds number of 3000. Turbulence is enhanced by increasing velocity and tube diameter but decreasing viscosity. (REF. 2, p. 775 ff)

429. A. The first heart sound occurs at the beginning of ventricular systole and is heard best over the apex of the heart. It consists of irregular vibrations, mainly of a frequency of 30–45 per second. These vibrations originate in three ways: (1) vibration from the AV valve during and after closure; (2) vibrations from the eddies in the stream of blood ejected through the aortic orifices as outflow into the wider sinuses of Valsalva; and (3) from vibrations in the cardiac muscle fibers themselves, generated as the heart muscles develop tension from a state of relaxation. In all, the main vibrations develop toward the end of the QRS at the beginning of ventricular contraction. (REF. 2, p. 840 ff)

430. E. The cardiac reserve includes any increase in cardiac performance no matter how it is achieved. (REF. 1, p. 312 ff)

431. D. In the untrained individual, the cardiac output increases during moderate exercise as a result of an increase in heart rate, but the stroke volume remains constant. (REF. 1, pp. 222, 1016)

432. E. Extrinsic control of ventricular function involves the neural and chemical stimuli to the heart. The neural control operates through efferents of the ANS, which ensure that the force of each contraction may be increased or decreased by alteration in the frequency and/or intensity of these impulses. In addition, the levels of O_2, CO_2, and catecholamines (as well as other neurochemical agents) in the blood constitute a second mode of cardiac extrinsic control. Myocardial hypoxia and hypercapnia depress contractility, while raising the catecholamine blood levels will result in an increase in cardiac inotropy. (REF. 2, p. 972 ff)

433. B. An increase in capillary permeability will allow a movement of fluid out of the vascular system into the body tissues. This will cause edema and a decreased plasma volume. (REF. 2, p. 865 ff)

434. D. The blood content of the capillaries is surprisingly small, in the range of 300 mL. (REF. 1, p. 218 ff)

435. E. Ventricular function curves express the functional ability of the heart at a time when the data for such curves were determined. Thus, the curves represent a relationship between right atrial pressure at the input of the heart and cardiac output from the left ventricle. Each curve indicates that, at any given right atrial pressure (venous return), the output increases in response to such variation in atrial filling or pressure. The degree of change depends on the conditional state of the heart. When conditions do change, as in failure (curve B), the ventricular function moves to a new output-pressure (or work-length) curve. Thus, the functional ability moved from curve A to curve B. Any successful therapeutic maneuver, as with digitalis administration, will shift the cardiac function to a new curve directed upward and to the left. Since the heart can shift from one function curve to another, the physiologic (and hence clinical) functional activity of the heart can be portrayed by a family of curves (i.e., a family of Starling law curves), with each curve corresponding to a specific set of circumstances. (REF. 2, p. 834 ff)

436. D. If we consider all adults, including older people, the average output is close to 5.0 L/min. (REF. 1, p. 272)

437. B. As a result of the elastic properties of the blood vascular wall, the size of the vessel lumen is a function of the forces applied to it. In the normal system, alteration in the dimension of most blood vessels along their longitudinal axis is only slight. However, changes in vessel circumference commonly occur in vivo, and this important relationship between pressure, wall tension, and the diameter of the elastic tube was expressed by LaPlace as $T = Pr$. The pressure, P, which is physiologically relevant, is the force acting radially on each unit area of the wall and includes the intraluminal pressure (P_I), as well as the pressure which is affecting the external surface of the vessel (P_E). This transmural pressure (P_T), which is exerted on the vessel wall, is the difference between internal and external pressure ($P_T = P_I - P_E$). This relationship usually operates in blood vessels with very thin walls and when wall thickness is increased, as with age or disease; then the pressure-tension relationship becomes extremely complex and clinically relevant. (REF. 2, p. 879 ff)

438. A. The parallel arrangement of most of the major vascular beds allows adjustments of blood flow through these beds independent of cardiac output. (REF. 2, p. 911 ff)

439. A. Transmembrane potential depends on the chemical and electrostatic forces across the cell membrane. This is due principally to the distribution of ions such as sodium, potassium, and calcium across the cell membrane. The difference in the permeability of the membrane to these ions results in a resting potential for that cell membrane. Changes in cell membrane permeability to these ions gives rise to an action potential due to the increased rate of passage of ions across the cell membrane. (REF. 3, pp. 398–399)

440. C. A sudden increase in the conductance (g_{Na}) of sodium ions results in an action potential, the magnitude of which varies linearly with the logarithm of the extracellular concentration of sodium. Therefore, when concentration of sodium falls to about 20 mM (140 mM is normal) the cell is no longer excitable. (REF. 3, p. 401)

441. E. The resting cell membrane is relatively permeable to potassium ions but less so to sodium ions. However, negatively charged intracellular proteins, which are not free to diffuse out of the cell, generate an electronegative charge when potassium ions leak out. Potassium ions continue to leak out until an electrostatic force counters the chemical force, based on concentration gradient. This electronegative force is usually −90 mV, negative with respect to the outside of the cell. Since the cell membrane is more permeable to potassium than sodium ions, the conductance of potassium (g_K) is approximately 100 times that of sodium (g_{Na}). (REF. 3, pp. 400–401)

442. C. Gating of the sodium channel is a voltage-dependent phenomenon in that the potential at which sodium channels swing open varies from channel to channel within the cell. As V_m becomes less negative more sodium channels open until the threshold value of −65 mV is reached, at which point, virtually all the sodium-fast channels open and rapid depolarization results in an action potential. (REF. 3, pp. 402–403)

443. A. Cardiac conduction is based on the simple principle that current, like a fluid, will flow from regions of higher potential to regions of lower potential. Propagation of the wave of excitation is due to the electrical conductivity of the electrolytes on either side of the membrane. Propagation is carried out by the inward flux of sodium ions at a particular site called the depolarized zone. This localized depolarization stimulates sodium conductance in a region adjacent to it. This process repeats itself over and over generating a wave of depolarization. (REF. 3, pp. 406–407)

444. C. Conduction of the fast response depends on factors affecting local depolarization. The greater the potential difference across a membrane the greater the action potential amplitude, the greater the magnitude of current, the more rapid the conduction. These factors are dependent on the rate of change potential (dV_m/dt) which reflects the rate of opening of the m gates or the fast sodium channels. (REF. 3, pp. 406–407)

445. C. The threshold for h gates is about −35 mV. The conduction characteristics are different than those of m gates. Conduction is usually in one direction; however, in bidirectional conduction, the velocity in one direction may be slower than for the other direction. Slow conduction responses are more likely to be blocked and cannot be conducted at as rapid repetition rates due to their slow refractory period. (REF. 3, p. 408)

446. A. Complete or third degree block is a condition in which the atria and ventricles are beating independently of each other. As a result, the P-R intervals vary and in fact are not related to each other. The ventricular complexes are not driven or initiated by atrial complexes. The ventricles beats are initiated by some pace maker at or below the AV nodal junction. (REF. 1, p 198 ff)

447. B. Automaticity is the ability of the heart to initiate its own heartbeat independent of nervous pathways. It does this by virtue of intrinsic cardiac tissue capable of initiating beats under circumstances where higher regions of rhythmicity are blocked or destroyed. At least some cells with this ability can be found in all four chambers of the heart. (REF. 3, pp. 409–410)

448. E. The SA nodal cells responsible for pacemaker activity are characteristic of slow response cells. They have a less negative resting potential (-65 mV). At this potential the h gates are closed, thereby deactivating the sparse sodium-fast channel in the nodal cells. (REF. 3, pp. 410–411)

449. A. Phase 2 is absent in action potentials generated by pacemaker cells of the SA node. Frequency of discharge of pacemaker cells is affected by the threshold potential, resting potential, and the rate of depolarization during phase 4. During phase 4 there is a slow and gradual depolarization until threshold is reached when an action potential is triggered. Any change in any one of these factors will affect the discharge frequency of the cell. (REF. 3, pp. 410–411)

450. C. There is an influx of sodium ions, but this takes place through the slow channels which are unaffected by tetrodotoxin. The main ionic effect on phase 2 of the action potential is due to the influx of calcium ions due to an increase in g_{Ca}. Since the V_m goes to zero chemical forces precipitate an efflux of potassium ions out of the cell. Though this efflux of potassium ions tends to make the interior of the cell more negative, and hence aids in repolarization, the g_K is decreased during phase 2. This is opposite to that found in nervouse tissue where g_K increases upon depolarization. (REF. 3, pp. 404–405)

451. E. A necessary condition for reentry, ordered or random, is that somewhere along the conduction pathway there is a blockage of the impulse in one direction while impulse may travel in the other direction. This is called unidirectional block. A typical example of random reentry is fibrillation, where the pathway for the impulse changes. Generally, reentry is the reexcitation of a region through which an impulse has previously passed. (REF. 3, pp. 416–417)

452. A. Electrocardiography is a valuable means of determining the electrical activity of the heart, and by recording potentials from various locations on the body, the physician can determine orientation of the heart as well as relative size of the chambers. Rhythmic disturbances can be detected as well as the effects of drugs such as digitalis. However, it should be remembered that electrocardiography relays no information regarding the mechanical performance of the heart. Electrical activity may appear normal, yet mechanical activity may be dangerously inadequate or lost altogether, a phenomenon called electromechanical dissociation. (REF. 3, pp. 418–419)

453. E. Third-degree AV block is referred to as complete heart block. Typically, these sites of complete blockage are distal to the bundle of His. There is complete mechanical dissociation between the atria and the ventricles. Consequently, the QRS complex has no fixed relationship to the P wave. With the reduced output associated with third-degree AV block, insufficient cerebral blood flow generally results in so-called Stokes-Adams attacks. (REF. 3, pp. 426–427)

454. A. The term hematocrit is used to express the volume of packed red cells as a percent of total blood volume. (REF. 1, p. 206)

455. E. Chemotaxis is a term used to describe the movement of leukocytes toward certain substances. (REF. 1, p. 53)

456. B. Pulse pressure is the difference between systolic and diastolic pressure. The increasing pulse pressure from the aorta to the peripheral arteries peaking in the femoral and saphenous arteries is a result of the increasing systolic and decreasing diastolic pressures. This is the result of wave reflection. (Note: the mean blood pressure, however, decreases steadily from the aorta toward the periphery.) (REF. 1, p. 225)

457. B. When the right atrial pressure rises above its normal value of 0 mm Hg, it begins to back up blood in the large veins and, hence, to open them up. The heart pumps into the arteries any excess blood that attempts to accumulate in the right atrium, and output is increased (Starling effect). (REF. 1, p. 221 ff)

458. B. In the early stages of essential hypertension the cardiac output is elevated, and as the blood volume increases, it in turn causes a further increase in cardiac output. This results in too much blood flow through the tissues, causing an increase in the total peripheral resistance. Long-term regulatory mechanisms attempt a return of cardiac output to its normal value. Thus, human essential hypertension may represent a vascular response to a high cardiac output that has led to an increased peripheral resistance and ultimately to a decrease in cardiac output. (REF. 1, p. 264 ff)

459. D. The fluid retention seen in chronic congestive heart failure results in an increased capillary hydrostatic pressure. (REF. 1, p. 306 ff)

460. D. In a patient with left bundle branch block, the second heart sound will likely be split because the aortic valve closure will be delayed or compared to the semilunar valve closure. (REF. 1, pp. 190, 321 ff)

461. C. A high resting heart rate with a high central venous pressure is often caused by the direct shunting of large amounts of blood from the arterial to the venous circulation. (REF. 2, p. 834 ff)

462. A. If stroke volume is increased with no change in heart rate, the cardiac output will increase. If there is no associated change in peripheral resistance, both pulse and mean pressure will increase. (REF. 1, p. 292 ff)

463. D. It can be shown that resistance to fluid flow through a tube is inversely proportioned to the fourth power of the radius of that tube. (REF. 1, p. 212 ff)

464. B. The correct order of valvular activation in a normal cardiac cycle is: mitral valve closes, tricuspid valve closes, aortic valve opens, pulmonic valve opens, aortic valve closes, pulmonic valve closes, tricuspid valve opens, and mitral valve opens. (REF. 1, p. 154 ff)

465. C. Increased pCO_2 is a potent stimulus for local vasodilation. (REF. 2, p. 853)

466. B. When high pressures in the right atrium cause back pressure, extra blood occasionally is stored in the hepatic veins and sinuses. (REF. 1, p. 221 ff)

467. E. Initially, when venous pressure increases, cardiac output increases. This results in an increase in arterial pressure and, consequently, a rise in capillary pressure. However, due to intrinsic mechanisms, i.e., delayed compliance, capillary pressure begins to drop. (REF. 1, p. 211)

468. E. If two vessels are connected in parallel, their total resistance to block flows is less than the resistance of any single vessel. (REF. 1, p. 212 ff)

469. D. The average velocity of blood flow in the femoral artery, or in any arterial vessel, is related in an inverse fashion to the cross-sectional area of that vessel. (REF. 1, p. 218 ff)

470. A. Bernoulli's principle states that pressure will decrease in smaller tubes. (REF. 3, p. 138 ff)

471. A. Homeometric autoregulation associated with an increase in mean aortic pressure can result in an increase in stroke work, but no real positive inotropy. (REF. 2, p. 849 ff)

472. B. If arterial capacitance remains constant, pulse pressure will vary directly with stroke volume. When the heart rate is increased, stroke volume decreases because of a shorter filling time. (REF. 2, p. 834 ff)

473. B. Resistance is the impediment to blood flow in a vessel. Slight changes in the diameter of a vessel cause tremendous changes in its ability to conduct blood. (REF. 1, p. 211 ff)

474. B. In order to measure cardiac output by the dye-dilution technique, the concentration of dye in arterial blood must be measured continuously. The cardiac output is then calculated by dividing the total amount of dye passing the monitoring point by mean dye concentration seen at this point. (REF. 2, p. 834 ff)

475. A. The midline vasodilator centers in the medulla have an inhibitory effect on the more lateral vasoconstrictor areas. The vascular effects are mediated through the sympathetic nervous system. (REF. 2, p. 972 ff)

476. A. Cardiac output (L/min) divided by the heart rate (beats/min) equals the mean stroke volume. (REF. 2, p. 972 ff)

477. E. It is the mean concentration of oxygen in the venous and arterial bloods that must be used in the calculations. (REF. 1, p. 284 ff)

478. A. The reactions to decreased pressure in the carotid sinus will be all those that tend to negate this pressure decrease. Vagal impulses are decreased to result in an increase in heart rate. (REF. 2, p. 972 ff)

479. C. Autoregulation of blood flow is shown by a vascular bed that maintains a constant blood flow over a range of changing blood pressures. (REF. 1, p. 260)

480. B. The smallest blood vessels have an inherent myogenic automaticity that controls local blood flow. This type of control is most important in the vascular beds of muscle. (REF. 2, p. 220 ff)

481. D. The second heart sound is due to the closing of the aortic valves. The closure of these valves also gives rise to the dicrotic notch of the arterial blood pressure. (REF. 2, p. 840 ff)

482. B. There is increased venous return during exercise due to increased work by skeletal muscle. This increased venous return does not appear to increase stroke volume. (REF. 2, p. 833 ff)

483. C. The total cross-sectional area of the capillaries is 2500 cm², compared to the second highest, venules, which is only 250 cm². (REF. 1, p. 219)

484. A. $\dfrac{\text{Total counts albumin injected}}{\text{Counts/mL albumin at equilibrium}} = \text{mL plasma}$
(REF. 3, p. 542 ff)

485. A. Reducing the pressure presented to the carotid baroreceptors results in a increase in peripheral resistance to return blood pressure towards normal. (REF. 2, p. 972 ff)

486. D. Baroreceptor activation slows the heart rate by increasing vagal parasympathetic activity and decreasing sympathetic stimulation of cardiac tissues. (REF. 2, p. 972 ff)

487. A. The cardiac output increases approximately in proportion to the surface area of the body or the cardiac output per square meter of body surface area. (REF. 1, p. 272)

488. E. The normal functional mean capillary pressure is about 17 mm Hg, well below that present in the larger arterial vessels. (REF. 1, p. 349)

489. D. Stretch receptors in the atria elicit a stretch reflex through the vagus nerves to the medulla of the brain. Efferent signals are transmitted back through both vagal and sympathetic nerves to increase heart rate and strength of contraction. This reflex helps prevent excess accumulation of blood in the veins, atria, and pulmonary circulation. This reflex is called the "Bainbridge reflex." (REF. 1, p. 250)

490. A. The only means by which impulses can ordinarily pass from the atria into the ventricles is through the bundle of His. Conditions such as ischemia of the AV junctional fibers, compression of the AV bundle, or inflammation of it can either decrease the rate of conduction of the impulses through the bundle or totally block the impulse. During such events, the interval between atrial and ventricular contractions is increased. (REF. 1, p. 198 ff)

491. D. In humans, the splenic capsule is nonmuscular. (REF. 1, p. 343)

492. B. The arterial pulse usually disappears in the arterioles. The term venous pulse is used to describe the pressure changes in the large veins near the heart that are due to changes occurring in the heart and adjacent large arteries. (REF. 1, p. 218 ff)

493. B. Bleeding time is the time necessary for a small nick to stop bleeding (2–3 minutes). This measurement is normal in hemophilia because the cessation of bleeding is due to the plugging of the wound by platelets which is normal in hemophilia. In hemophilia the coagulation time (but not bleeding time) is usually prolonged because the thromboplastin activity is abnormal. (REF. 1, p. 85 ff)

494. D. The primary pacemaker of the heart is the SA node. Activity originating from any other site represents an abnormal situation, usually such "secondary" pacemaker activity resides within the Purkinje system. (REF. 2, p. 799)

495. B. Tetany can occur in skeletal muscle because several contractile events can be summated. The long electrical refractory period of cardiac muscle precludes this possibility. (REF. 1, p. 154)

496. B. Aortic flow velocity reaches a maximum during the early rapid phase of ventricular ejection and then decreases. (REF. 1, p. 208 ff)

497. C. If large regions of abnormally low V/Q ratio do exist, there will be, in effect, a shunt of blood past the oxygenation mechanism of the lungs. This will result in both cyanosis and systemic arterial blood hypoxemia. (REF. 2, p. 1022 ff)

498. E. The blood pressure found in the main pulmonary artery is not inversely proportional to lung tissue or the cardiac output. Also, alveolar hypoxia does not result in vasodilation, nor is this pressure always high enough to perfuse the top of the human lungs. (REF. 2, p. 1012 ff)

499. A. The area of the pressure-volume diagram is a very close approximator of cardiac work. (REF. 2, p. 834 ff)

500. B. Coronary vasodilation is the normal response to cardiac sympathetic stimulation. (REF. 2, p. 944 ff)

501. C. The myogenic theory of autoregulation of blood flow proposes that flow to a tissue bed is kept constant when pressure increases because of an inherent adjustment of smooth muscle to stretch. (REF. 2, p. 220 ff)

502. D. Venoconstriction adds to a redistribution of blood and increased cardiac output during exercise. (REF. 2, p. 834 ff)

503. B. Flow through a rigid tube is related inversely to the first power of the length of that tube. (REF. 2, p. 849 ff)

504. B. In normal ranges as the diastolic volume of the heart increases the force and the rate of contraction of cardiac muscle fibers will increase. (REF. 2, p. 834 ff)

505. B. A rigid aorta will accommodate less blood during ejection, thereby raising the resistance to movement of blood out of the heart. (REF. 1, p. 269 ff)

506. B. A patient with the signs listed has normal function except for the increased blood pressure. This indicates hypertension. (REF. 1, pp. 264, 270 ff)

507. A. Any condition that increases blood pooling contributes to congestive heart failure. This would include decreased arterial pressure, increased serum sodium, and increased tissue hydrostatic pressure. (REF. 1, p. 305 ff)

508. C. During exercise, the greatest portion of additional O_2 supplied to working tissues comes from an increase in arteriovenous oxygen difference. (REF. 2, p. 1584 ff)

509. B. With a cardiac output reduction to one third of normal and an arterial pressure of 70/40, the venous pressure is likely to be increased. (REF. 1, p. 279 ff)

510. D. Poiseuille's equation states that flow through a cylindrical tube is proportional to the pressure gradient and the fourth power of the radius, and inversely proportional to the length of the tube and fluid viscosity. (REF. 2, p. 778 ff)

511. B. Purkinje fibers are larger than the normal ventricular muscle fibers and they transmit impulses at a velocity of 1.5–4.0 m/s. This is a velocity of about six times that in the usual cardiac muscle and 300 times that in the junctional fibers of the AV node. (REF. 2, p. 799 ff)

512. D. The findings listed associated with edematous lower extremities are indicative of an obstruction to venous return from the lower portion of the body. (REF. 2, p. 875 ff)

513. D. $CO = \dfrac{mL/min \ of \ O_2 \ used}{AV \ O_2 \ mL/min} = 5.51 \ L/min.$
(REF. 2, p. 229 ff)

514. D. Stroke volume $= \dfrac{CO \ mL/min}{Heart \ beats/min} = 55 \ mL/beat.$
(REF. 2, p. 229 ff)

515. A. CO = Heart rate × Stroke volume = 10.5 L/min. (REF. 2, p. 229 ff)

516. A.

517. B.

518. C.

519. D.

520. D. The mean blood pressure decreases as the blood moves from the aorta into capillaries. Because of the inverse relationship between flow velocity and cross-sectional area, the blood flows much more slowly in the distal arteries and especially in the arterioles with the lowest velocity in the capillaries. This permits the greatest fluid, nutrients, and waste product exchange. The resistance to flow is determined largely by the blood viscosity and radius. (Hence, the viscosity in the capillaries is less than in the arterioles.) The greatest resistance to flow is in the arterioles. (REF. 1, p. 219 ff)

521. C. Approximately 0.16 seconds after the onset of the P wave, the QRS waves appear as a result of depolarization of the ventricles. (REF. 1, p. 154)

522. E. The atrial C wave occurs because of the bulging of the AV valves toward the atria. This comes about because of the increasing pressure in the ventricles and pulling on the atrial muscle where it is attached to the ventricular muscle by the contracting ventricles. (REF. 2, p. 834 ff)

523. F. As soon as systole is over and the ventricular pressures fall again to their low diastolic values, the high pressures in the atria immediately push the AV valves open and allow blood to flow rapidly into the ventricles. (REF. 2, p. 834 ff)

524. D. The T wave represents the stage of repolarization of the ventricles at the time the ventricular muscles begin to relax. (REF. 2, p. 796)

525. B. At the end of systole, ventricular relaxation begins suddenly, allowing the intraventricular pressures in the large arteries to immediately push blood back toward the ventricles, which snaps the aortic and pulmonary valves closed. (REF. 2, p. 834 ff)

526. A. When ventricular contraction is over the AV valves open, allowing blood to flow rapidly into the ventricles. (REF. 2, p. 834 ff)

527. A. The first heart sound is largely due to closure of the atrioventricular valves and coincides with the period of isometric contraction. (REF. 2, p. 834 ff)

528. D. The Q wave of the ECG immediately precedes isovolumic contraction. (REF. 2, p. 834 ff)

529. E. The T wave of the ECG occurs during the period of ventricular ejection. (REF. 2, p. 834 ff)

530. C. The P wave of the ECG is due to depolarization of the atria and immediately precedes atrial contraction. (REF. 2, p. 796)

531. B. During the period of isometric relaxation the second heart sound can be heard. (REF. 2, p. 834 ff)

532. B. Mean arterial pressure is decreased by aortic regurgitation and mitral stenosis. Aortic regurgitation is caused by the return of blood from the elastic aorta through the defective valve during diastole. Mitral stenosis is an obstruction to the flow through the orifice so that blood must be forced from the atrium during diastole. Both result in arterial pressure. (REF. 1, p. 319 ff)

533. A. In aortic regurgitation, return of blood from the aorta through the defective valve causes increased left atrial pressure. (In mitral valve stenosis, there is an obstruction to blood flow from the atrium, also resulting in increased atrial pressure.) (REF. 1, p. 319 ff)

534. B. In aortic regurgitation, return of the blood through the defective valve causes a decreased volume of blood pumped into systemic circulation, thus, a decreased cardiac output is caused. In mitral stenosis, there is a decreased volume of blood pumped into the left ventricle, resulting in decreased cardiac output. (REF. 1, p. 319 ff)

535. A. In aortic regurgitation, there is increased blood return to the left ventricle by the defective valve with resulting increased left ventricular volume. In mitral stenosis, there is decreased filling of the left ventricle resulting in increased left atrial blood volume. (REF. 1, p. 319 ff)

536. A. In aortic regurgitation, there is increased return of blood through the defective valve to increase left ventricular volume. In mitral stenosis, there is a decreased filling of the left ventricle by the defective valve, resulting in decreased left ventricular volume. (REF. 1, p. 319 ff)

537. A. In aortic regurgitation, there is decreased cardiac output, as is true in mitral stenosis, due to defective valves. This results in decreased peripheral flow which envokes a compensatory reflex. Renal output slows down until blood volume increases. (REF. 1, p. 319 ff)

538. A. It is the first electrical event recorded on the ECG and represents atrial depolarization. (REF. 2, p. 796 ff)

539. H. Represents part of the ventricular depolarization the wavefront goes down the septum. (REF. 2, p. 796 ff)

540. J. Represents the major vector of ventricular depolarization. (REF. 2, p. 796 ff)

541. I. The terminal portion of ventricular depolarization. (REF. 2, p. 796 ff)

542. C. A ventricular muscle repolarization (phase 3 of the action potential). (REF. 2, p. 796 ff)

543. B. May be the representation of Purkinje fiber repolarization and/or afterpotentials. (REF. 2, p. 796 ff)

544. E. The time during which the impulse activates the atria and passes through the AV node and specialized conducting system. (REF. 2, p. 796 ff)

545. D. The complete or total representation of ventricular muscle depolarization. (REF. 2, p. 796 ff)

546. F. Terminal ventricular depolarization to onset of repolarization is temporally equal to the phase 2 of ventricular muscle action potentials. (REF. 2, p. 796 ff)

547. G. The entire period of time of ventricular repolarization includes phases 2 and 3 of ventricular muscle action potentials. (REF. 2, p. 796 ff)

Summary of answers 538–547: The ECG deflections comprise four positive and three negative waves. Some of these waves are termed complexes when they are combined and all are descriptive, indicating some specific physiologic event(s) occurring in the heart. The waves, the various intervals and segments, provide a way of measuring the direction of a specific event. Both the contours and polarity of the complexes recorded from each lead of the ECG depend on the orientation of the excitation fronts of the heart. (REF. 2, p. 796 ff)

7 Endocrine Physiology

DIRECTIONS (Questions 548–606): For each of the questions or incomplete statements below, **one** or **more** of the answers or completions given is correct. Select
    A if only *1, 2, and 3* are correct
    B if only *1 and 3* are correct
    C if only *2 and 4* are correct
    D if only *4* is correct
    E if all are correct

548. Which of the following is(are) true regarding progesterone?
1. Secreted by the corpus luteum
2. Produced by the ovaries
3. Secreted by the placenta
4. Suppresses fertility

549. Estrogens are produced by the
1. placenta
2. Leydig cells of the testes
3. ovary
4. adrenal cortex

550. Spermatogenesis requires
1. adrenocorticotropic hormone (ACTH)
2. androgens
3. thyroid-stimulating hormone (TSH)
4. follicle-stimulating hormone (FSH)

551. After fertilization of the ovum the
1. corpus luteum is supported by human chorionic gonad-otropin (HCG)
2. conceptus begins to divide in the fallopian tube
3. uterine engorgement is maintained by luteal progester-one for several days
4. corpus luteum finally fails after about 2 weeks

552. Secretion of glucocorticosteroids is
1. increased by ACTH
2. increased by stress
3. decreased if exogenous glucocorticosteroids are given
4. usually not related to CNS activity

553. The physiologic action(s) of estrogen on vagina include:
1. deposition of glycogen
2. mucosal thickening
3. epithelial cornification
4. increased mitosis; growth and differentiation of muco-sal layers

554. Diabetes mellitus is characterized by
1. polydipsia
2. polyuria
3. polyphagia
4. glycosuria

555. Pituitary secretion
1. shows a negative feedback relationship with its target organs
2. will increase as long as target gland secretion is high
3. will increase if levels of circulating target gland secre-tions are low
4. if increased will usually cause target gland atrophy

556. The thyroid hormones
1. act on discrete target organs
2. do not act immediately on their target cell following their release
3. act on discrete organ systems
4. are not necessary for maintenance of life

Directions Summarized				
A	**B**	**C**	**D**	**E**
1,2,3	1,3	2,4	4	All are
only	only	only	only	correct

557. Anterior pituitary secretion
    1. depends on a direct neural connection between the hypothalamus and the anterior pituitary
    2. of only one pituitary hormone is signaled by each hypothalamic-releasing factor
    3. is independent of CNS control from above the level of the hypothalamus
    4. depends on the release of hypothalamic-releasing factors into the primary capillary plexus in median eminence

558. The adrenal hormone cortisol plays a major role in
    1. sodium chloride retention
    2. potassium excretion
    3. hydrogen ion excretion
    4. maintaining blood sugar levels

559. Excessive circulating levels of growth hormone are associated with
    1. decrease in the body glycogen stores
    2. mobilization of fatty acids from adipose tissue
    3. increased concentration of amino acids in blood
    4. delayed bone maturation

560. Secretion of adrenocorticotropic hormone (ACTH)
    1. supports the adrenal medulla
    2. is released into the anterior pituitary circulation from neurons with cell bodies in the hypothalamus
    3. in excess causes hypertrophy of the adrenal medulla
    4. does not evoke the secretion of the adrenal hormones cortisol and corticosterone

561. During a normal menstrual cycle
1. the corpus luteum is controlled by the FSH concentration in the blood
2. the rising concentration of estrogen secreted by the ripening follicle triggers the luteinizing hormone (LH) peak
3. ovulation will occur exactly 14 days after the last vaginal bleeding
4. the lifetime of the corpus luteum is about 10 days

562. Luteinizing hormone (LH) in males
1. stimulates secretion of testosterone
2. is necessary for formation of adult sperm
3. has a trophic effect on the Leydig cells
4. acts on the seminiferous tubules to encourage sperm formation

563. Luteinizing hormone
1. supports the corpus luteum
2. increases blood flow to the ovary
3. triggers ovulation
4. causes the secretion of progesterone by the follicle, which in turn causes the release of an enzyme that breaks open the follicle

564. A large dose of exogenous insulin would result in
1. increased movement of glucose from the blood into many cells of the body
2. decreased secretory activity of the alpha cells of the pancreas
3. increased secretion of glucagon
4. increased secretion of beta cells of the pancreas

565. Thyroid-stimulating hormone (TSH)
1. is necessary for adequate release of thyroid hormones
2. promotes iodine trapping
3. increases the thyroglobulin iodination
4. increases the movement of thyroglobulin into follicle cells

Directions Summarized				
A	**B**	**C**	**D**	**E**
1,2,3	1,3	2,4	4	All are
only	only	only	only	correct

566. Thyrocalcitonin
1. is secreted by the parathyroid
2. is secreted by the thyroid
3. increases Ca^{2+} absorption by the stomach
4. decreases mobilization of Ca^{2+} from bone

567. The specific effect of thyroid hormones on the basal metabolic rate (BMR)
1. results in decreased O_2 consumption when thyroid hormone levels are increased
2. persists for several days after the disappearance of a single dose of T_4
3. is almost completely due to the stimulation of metabolism of skeletal muscle
4. requires the stimulation of the sodium pump by the thyroid hormones

568. Oxytocin
1. is secreted by the posterior pituitary (neurohypophysis)
2. secretion is associated with neural activity
3. is secreted directly from nerve endings
4. is secreted by the anterior pituitary (adenohypophysis)

569. The ovarian follicles
1. secrete estrogens
2. secrete luteinizing hormone
3. secrete progestins
4. supply only the gametes for reproduction

570. Insulin
1. is necessary for glycogenolysis
2. increases fatty acid synthesis
3. has no effect on fatty acid metabolism
4. is necessary for the storage of foodstuffs during a meal

571. After the first 5 months of pregnancy
1. uterine engorgement is maintained by placental estrogen and progesterone
2. uterine engorgement is maintained by follicular estrogen and progesterone
3. high levels of estrogen inhibit further ovulation
4. FSH maintains a normal rate of follicular maturation

572. The action of parathormone on the kidney tends to
1. increase calcium phosphate precipitation in the kidney
2. effectively increase the deposition of bone mineral
3. elevate serum phosphate
4. diminish the possibility of hyperphosphatemia

573. Glucocorticoid hormones from the adrenal cortex
1. increase glucose formation in the liver
2. increase the concentrations of enzymes necessary for glucose synthesis
3. mobilize amino acids from muscle
4. increase the movement of amino acids from the blood into most cells of the body

574. Parathyroid hormone
1. decreases Ca^{2+} release from bone
2. increases Ca^{2+} release from bone
3. increases urinary excretion of Ca^{2+}
4. decreases urinary excretion of Ca^{2+}

575. Serum cholesterol is elevated by
1. reabsorption and recirculation of bile acids
2. obstruction of the common bile duct
3. ingestion of dietary cholesterol
4. ingestion of dietary fatty acids

576. Abnormally low concentrations of estrogen result in
1. increased height
2. lack of or very late appearance of pubic hair
3. decreased deposition of fat in the subcutaneous area
4. increased retention of Na^+ and H_2O

Directions Summarized				
A	**B**	**C**	**D**	**E**
1,2,3	1,3	2,4	4	All are
only	only	only	only	correct

577. The function of the endocrine system includes which of the following?
    1. Regulation of growth and maturation
    2. Behavior of the organism and its reproduction
    3. Regulation of metabolic substrates and mineral flow for maintenance of chemical homeostasis
    4. Body weight and size

578. Which of the following is considered true regarding the endocrine system versus the neural system?
    1. Both transmit signals that are highly focused and localized
    2. Both transmit signals that are widespread and diverse
    3. Each is basically a system for signaling in a stimulus-response manner
    4. The two systems differ only in their modes for transmission of chemical signals

579. The neurocrine mechanism may be characterized by
    1. autocrine effects
    2. a mechanism of transmission that includes the circulatory system
    3. the relatively short distance involved during the transmission of messenger molecules
    4. a concept of a "neurohormone"

580. According to the neuroendocrine system concept, a messenger molecule may function as a(an)
    1. paracrine hormone only
    2. paracrine and autocrine hormone
    3. endocrine hormone only
    4. endocrine and neurotransmitter

581. Which of the following is considered true regarding hormone synthesis?
1. Nuclear posttranscriptional modification of the primary gene transcript includes excision of introns, splicing of exons, capping of the 5′ end, and addition of poly-A at the 3′ end
2. Mature messenger RNA directs the synthesis of a preprohormone peptide sequence on ribosomes, after which the N-terminal signal is removed, and the resultant peptide is transferred vectorially into the endoplasmic reticulum
3. Final cleavage of the prohormone takes place in the secretory granules and the Golgi apparatus, where it is then stored in granules ready for secretion by exocytosis
4. Copeptide synthesis and secretion parallel hormone synthesis and release

582. Catecholamine and protein hormones are characterized as having the following properties:
1. simple compartmentalization with subsequent transfer of the free cytosolic form through the plasma membrane
2. utilize secretory granules for storage and transport to cell membrane for exocytosis into the extracellular space
3. mobilized as a result of stimuli that lower intracellular cAMP with a concomitant rise in cytosolic calcium ions
4. mobilized as a result of stimuli that usually raise cytosolic calcium ion levels along with increased levels of intracellular cAMP

583. One mode of hormone synthesis and release involves the modification of hormone X from cell X by adjacent cell Y to produce hormone Y. This mode is typical of
1. vitamin D hormone synthesis
2. angiotensin hormone synthesis
3. adrenalin hormone synthesis
4. estrogen hormone synthesis

Directions Summarized				
A	**B**	**C**	**D**	**E**
1,2,3	1,3	2,4	4	All are
only	only	only	only	correct

584. Chronotropic control is a general mechanism for governing hormone secretion. Typical examples of chronotropic control are
    1. menstrual rhythm
    2. seasonal rhythm
    3. developmental rhythm
    4. sleep-wake cycle

585. Assuming that "product" is a function of hormone secretion, which of the following is(are) associated with the principle of negative feedback control of hormone secretion?
    1. Increase in hormone secretion stimulates a greater output of product from a target cell
    2. Increased product from target cell suppresses further hormone secretion
    3. Hormone secretion may be stimulated by a deficit in product serum levels
    4. Hormone secretion may be inhibited by a deficit in product serum levels

586. Receptors for steroid hormones
    1. are located in the plasma membrane
    2. are located in the cytosol
    3. are located in the nucleus
    4. are relatively large protein molecules ranging in number from 2000 to 100,000 per cell

587. With increasing concentration of hormone in solution with a fixed number of cells, the amount of bound hormone
1. increases until 100% of receptor sites are occupied
2. increases until bound hormone equals the K value for the association, assuming free hormone concentration approaches infinity
3. to free hormone ratio approaches zero, assuming free hormone concentration approaches infinity
4. to free hormone ratio approaches 1, assuming free hormone concentration approaches infinity

588. In a cell where receptor binding is rate limiting
1. the maximal responsiveness of the cell to hormones is increased with an increase in receptor site number
2. increase in receptor site number enhances the cell's sensitivity
3. increase in receptor affinity increases the cell's sensitivity
4. sustained levels of excess hormone concentration result in increased number of receptor sites

589. Which of the following is NOT a system for coupling hormone recognition to hormone action?
1. Adenylate cyclase-cAMP system
2. Membrane phospholipid system
3. Calcium-calmodulin system
4. Lysozymal degradation system

590. Which is NOT a feature of the adenylate cyclase-cAMP system?
1. Conversion of GTP to GMP concomitant conversion of ATP to cAMP
2. Conversion of GTP to GDP with concomitant conversion of Mg-ATP to cAMP
3. Activation of regulatory unit by GDP activates catalytic unit
4. Activation of regulatory unit by GTP activates catalytic unit

		Directions Summarized		
A	**B**	**C**	**D**	**E**
1,2,3	1,3	2,4	4	All are
only	only	only	only	correct

591. Which of the following may be said regarding the calcium-calmodulin system for the transduction of hormonal signals?
    1. Requires specific plasma membrane proteins essential to creating calcium channels upon binding of hormone to receptor sites
    2. Requires calcium-specific binding protein
    3. Serves to deactivate as well as activate various enzyme and metabolic pathways
    4. Serves only to regulate intracellular calcium ion concentrations

592. Steroid hormones
    1. are specific for cytoplasmic receptors
    2. effects are fairly rapid, which is typical of plasma membrane response mechanisms
    3. effects are fairly slow due to their mode of transcriptional and translational regulation
    4. are specific for membrane-bound receptors

593. Which of the following is NOT a feature of the phospholipid (phosphotidyl) system of hormone action?
    1. Formation of phosphotidyl inositol-4,5-*bis*-phosphate
    2. Activation of a non-calcium-dependent protein kinase by the action of diacylglycerol
    3. Cleavage of phosphotidyl inositol-4,5-*bis*-phosphate by membrane-bound phospholipase
    4. Release of arachidonic acid by hydrolysis of diacylglycerols serves as substrate for the synthesis of calmodulin

594. Which of the following is a feature of (measurable) hormone responsivity
    1. Receptor number
    2. Hormone concentration
    3. Concentration of rate-limiting enzymes, cofactors, or substrates
    4. Minimal threshold concentration of hormone

595. Which of the following is(are) true regarding the dose response for the action of hormones?
    1. Response is generally not an all-or-none phenomenon
    2. Target cells require a minimal threshold concentration for measurable response
    3. Concentration of hormone required to elicit a half-measurable response defines the sensitivity of the cell
    4. Intrinsic basal-level activity may be observed independent of response and any added hormone

596. The extent of protein binding
    1. is low for all thyroid and steroid hormones (< 10%)
    2. is greater for protein hormones than thyroid/steroid hormones
    3. affects excretion rates and half-life
    4. renders hormone exit into interstitial fluid irreversible

597. Metabolic clearance rate (MCR)
    1. is the sum of all removal processes, such as urinary excretion, metabolic degradation, and target cell uptake
    2. is defined as the volume of plasma cleared per unit time
    3. is a measure of the efficiency with which hormones are removed from the plasma
    4. means clearance rate

598. Under ordinary basal metabolic conditions
    1. fat is a minor energy source
    2. fat provides over half the daily caloric requirements
    3. the respiratory quotient (RQ) for glucose is higher than for fat
    4. the RQ for fat is higher than for protein

		Directions Summarized		
A	**B**	**C**	**D**	**E**
1,2,3	1,3	2,4	4	All are
only	only	only	only	correct

599. In the postabsorptive state (overnight fast) for a normal adult
 1. plasma glucose levels are 60 to 115 mg/dL
 2. lactate is the predominant precursor for gluconeogenesis in maintaining blood glucose level
 3. the brain accounts for almost half of all glucose utilization
 4. the significant source of glucose in postabsorptive state is hepatic gluconeogenesis

600. In amino acid metabolism in an adult
 1. all 20 amino acids can be completely oxidized to CO_2 and water after deamination
 2. all 20 amino acids are required in the diet of infants
 3. all 20 amino acids give rise to ammonia by transamination or oxidative deamination during degradation
 4. all 20 amino acids are glucogenic

601. Regarding lipid metabolism
 1. chylomicrons are formed in the liver
 2. high-density lipoproteins (HDLs) are the major source of plasma triglycerides during overnight fast
 3. chylomicrons have a half-life of 2 hours
 4. low-density lipoproteins (LDLs) contain apoproteins specific for the receptor-mediated uptake of LDL by most tissues

602. The known hormone(s) released by the islet of Langerhans is(are)
 1. insulin
 2. glucagon
 3. somatostatin
 4. pancreatic polypeptide

603. The source of insulin is
1. paracrine cells
2. sigma cells
3. alpha cells
4. beta cells

604. Insulin secretion is
1. stimulated by blood glucose levels as low as 10 mg/dL
2. stimulated by fuel excess
3. inhibited by sulfonylurea compounds
4. inhibited by exercise

605. Once insulin is secreted
1. the majority is excreted unchanged in the urine
2. it circulates unbound
3. it has a plasma half-life of 30 to 40 minutes
4. it is degraded principally in the liver and kidney

606. The binding of insulin to receptor stimulates uptake of
1. glucose
2. amino acids
3. phosphate and potassium
4. magnesium

DIRECTIONS (Questions 607–617): Each of the questions or incomplete statements below is followed by five suggested answers or completions. Select the **one** that is **best** in each case.

---

607. Which of the following hormones is NOT secreted by the adenohypophysis?
A. ACTH
B. ADH
C. Growth hormone
D. TSH
E. FSH

608. The hormone with the shortest half-life in the bloodstream in the following list is
   A. insulin
   B. T_3
   C. ACTH
   D. epinephrine
   E. acetylcholine

609. Urinary 17-ketosteroids
   A. are not found in women
   B. reflect the total production of androgenic substances
   C. indicate the total production of sex hormones
   D. include testosterone
   E. are highly active androgens

610. Parathyroid hormone
   A. is released when serum Ca^{2+} is too high
   B. inactivates vitamin D
   C. is secreted if serum Ca^{2+} is too low
   D. works in the same direction as thyrocalcitonin
   E. depends on vitamin K for adequate activity

611. The development of normal lactation depends on adequate amounts of all of the following EXCEPT
   A. prolactin
   B. T_3 and T_4
   C. growth hormone
   D. androgens
   E. oxytocin

612. The hormone measured in urine to test for pregnancy is
   A. pituitary luteinizing hormone
   B. androgen
   C. progesterone
   D. chorionic gonadotropin
   E. follicle-stimulating hormone

613. If a patient presents with a very regular 29-day menstrual cycle, ovulation should occur on day
    A. 5
    B. 14
    C. 18
    D. 20
    E. 29

614. The hypothalamus is required for
    A. rhythmic breathing
    B. synthesis of anterior pituitary hormones
    C. blood pressure homeostasis
    D. perception of odor
    E. homeostasis of body temperature

615. During pregnancy the maximum rate of secretion of which of the following occurs during the first trimester?
    A. Chorionic gonadotropin
    B. Estrogen
    C. Pregnanediol
    D. Oxytocin
    E. Hydrocortisone

616. The secretion of testosterone by the testes
    A. occurs in the Sertoli cells
    B. is the responsibility of mature sperm cells
    C. is under control of the sympathetic innervation of the area
    D. reaches peak at about 20 years of age in the normal male
    E. none of the above

617. The primary stimulus for insulin secretion is increased
    A. epinephrine blood levels
    B. glucagon blood levels
    C. glucose blood levels
    D. stress
    E. water intake

DIRECTIONS (Questions 618–620): The group of questions below consists of a set of lettered components, followed by a list of numbered words or phrases. For each numbered word or phrase, select the one lettered component that is most closely associated with it. Each lettered component may be selected once, more than once, or not at all.

    A. Insulin
    B. Glucagon
    C. Both
    D. Neither

Encourages

618. Glycogenolysis

619. Metabolic actions of epinephrine

620. Gluconeogenesis

DIRECTIONS (Questions 621–625): The set of lettered items below is followed by a list of numbered words or phrases. For each numbered word or phrase select
    A if the item is associated with **A** only
    B if the item is associated with **B** only
    C if the item is associated with both **A** and **B**
    D if the item is associated with neither **A** nor **B**

    A. Estrogenic substances
    B. Androgenic substances
    C. Both
    D. Neither

621. Produced by the ovary

622. Produced by the testes

623. Produced by the adrenal cortex

624. Produced in measurable amounts throughout adult life

625. Anabolic action on muscle and bone

Explanatory Answers

548. E. All the statements listed are true regarding progesterone. (REF. 1, p. 976 ff)

549. E. Estrogens are known to be secreted by a variety of tissues. (REF. 1, p. 974)

550. C. Both androgens and FSH are required for the normal development and maturation of sperm. (REF. 1, p. 954 ff)

551. E. The extension of the life of the corpus luteum for 2 weeks by HCG is essential for the maintenance of the uterine wall during the early periods of pregnancy. The conceptus has begun to divide before it leaves the fallopian tube. Uterine engorgement is maintained by luteal progesterone, and after 2 weeks the corpus luteum degenerates. (REF. 1, p. 983 ff)

552. A. The secretion of glucocorticosteroids is directly proportional and arranged in a negative feedback loop with the secretion of ACTH. Further, these hormones will be released when CNS functions alter under stress. (REF. 1, p. 960 ff)

553. E. Estrogen causes mucosal thickening with cornification of the superficial vaginal epithelium. Mitotic figures appear in the basal layer of the vaginal mucus with growth and differentiation in the intermediate and superficial layers. In addition, there is a very heavy glycogen deposition in the intermediate and superficial layers, which may be as high as 3%. This glycogen deposition is very important in the maintenance of the low vaginal pH (4–5), which protects against infection. The low pH is a result of the vaginal bacteria which acts upon the glycogen breakdown products to produce lactic acid. (REF. 1, p. 974 ff)

554. E. As a result of the hyperglycemia in the diabetic, there is such a high amount of glucose in the glomerular filtrate that the transport maximum is exceeded and glucose spills into the urine. In an attempt to compensate for the glucose loss in the urine, there is an increase in appetite. The large output of glucose in the filtrate causes an osmotic hindrance to water reabsorption in the proximal tubules, which cannot be reabsorbed in the distal tubule. The result is osmotic diuresis. Furthermore, the copious urine flow results in dehydration and stimulates thirst, so that the diabetic is constantly thirsty and consumes large volumes of water. (REF. 1, p. 933)

555. B. Pituitary hormone secretion has a negative feedback relationship to many of its target organs. This means that if the target organ secretions are low, the pituitary trophic secretions will be high. (REF. 1, p. 879 ff)

556. C. The hormones of the thyroid do not have any discrete target organ or systems, and their effects are diffuse, being manifest throughout the body. Their effects are observed after a considerable lag or latency that may last for several days, indicating that these hormones appear to play a role in long-term functions rather than instant minute-to-minute regulation. Furthermore, these hormones appear to not be essential for maintaining life, albeit their lack certainly reduces the quality of life. (REF. 1, p. 876)

557. D. Adenohypophyseal secretions are stimulated by releasing factors formed in hypothalamic neurons and released into the primary capillary plexus in the median eminence. (REF. 1, p. 884 ff)

558. E. The adrenal cortex behaves as two different endocrine glands. It secretes the mineralocorticoid aldosterone which enhances excretion of sodium and hydrogen while retaining potassium. The cortex also secretes cortisol which helps to sustain blood sugar and blood pressure. (REF. 1, p. 909 ff)

559. C. Overproduction of growth hormones will result in a prominent increase in bone growth with delayed maturation. In addition, growth hormone will favor protein production over all other biochemical paths, and hence, a mobilization of fatty acids from adipose tissue for energy. (REF. 1, p. 887 ff)

560. B. ACTH will cause the secretion of cortisol and corticosterone from the adrenal cortex after its release from cells in the anterior pituitary. (REF. 1, p. 884)

561. C. During a normal menstrual cycle, the rising concentration of estrogen secreted by the ripening follicle triggers the LH peak, which causes ovulation. In the case where fertilization does not occur, the lifetime of the corpus luteum is about 10 days. (REF. 1, p. 969 ff)

562. A. The major role of luteinizing hormone in the male is the support of the Leydig cells and stimulation of testosterone secretion. However, LH is also needed to induce the Leydig cells to make the testosterone needed for spermatogenesis. (REF. 1, p. 963 ff)

563. E. Luteinizing hormone supports the ovary by increasing blood flow to that organ. In addition, this hormone triggers ovulation by increasing progesterone secretion, which in turn causes the release of an enzyme that breaks open the follicle. After ovulation, LH supports the continued existence of the ovum. (REF. 1, p. 968 ff)

564. B. A large dose of exogenous insulin would lower blood glucose levels by encouraging the movement of glucose into most of the cells of the body. This lowered glucose level would cause a reflex increased secretion of glucagon. (REF. 1, p. 924 ff)

565. E. Thyroid-stimulating hormone activates all of the thyroid functions listed. (REF. 1, p. 903 ff)

566. C. Thyrocalcitonin is secreted by the thyroid gland and is important in control of Ca^{2+} mobilization from bone. (REF. 1, p. 946 ff)

567. C. The effect of thyroid hormones on the basal metabolic rate persists for several days after a single dose of T_4, and is blocked by ovulation. (REF. 1, p. 900 ff)

568. A. Oxytocin is a hormone that is secreted from nerve endings found in the posterior pituitary gland. (REF. 1, p. 893)

569. B. The ovarian follicles supply both of the major classes of ovarian hormones (estrogens and progestins) in addition to supplying the gametes for reproduction. (REF. 1, p. 968 ff)

570. C. The major role of insulin is to facilitate the storage of foods during a meal. A portion of this activity is an increase in the synthesis of fatty acids as a preliminary to triglyceride formation. (REF. 1, pp. 823, 926 ff)

571. B. During the second trimester of pregnancy, inhibition of ovulation and uterine enlargement are maintained by high levels of estrogen and progesterone secreted by the placenta. (REF. 1, p. 987 ff)

572. D. The major function of parathormone on the kidney is to increase the excretion of phosphate ion. (REF. 1, p. 943)

573. A. The glucocorticoids increase the supply of glucose in the bloodstream through several mechanisms. The most important of these are increased gluconeogenesis, increased enzyme concentrations for glucose formation, and increased substrate by mobilizing amino acids primarily from muscle. (REF. 1, p. 914 ff)

574. C. Parathyroid hormone causes an increase in circulating levels of free Ca^{2+} by mobilizing Ca^{2+} from bone and decreasing excretion of Ca^{2+} by the kidney. (REF. 1, p. 948 ff)

575. E. All of the factors listed in the question will raise serum cholesterol. (REF. 1, p. 825)

576. A. Estrogen antagonizes growth hormones and encourages closure of the epiphyseal plate. The deposition of subcutaneous fat and development of pubic hair also require estrogen. Low estrogen secretion does not increase Na^+ and H_2O retention. (REF. 1, p. 974)

577. E. Nearly all aspects of an organism's development and maintenance, including the ones mentioned in the question, can be traced to effects associated with the endocrine system. (REF. 3, p. 819)

578. E. Advanced immunohistochemical observations suggests that the relationship between hormone secretion and neural function is so intimate that the distinction between the two systems is increasingly obscured. Peptides once thought of as classic endo-crine products have been found in such diverse places as the brain and neural tissue. (REF. 3, p. 819)

579. C. Neurons possess both autocrine and paracrine effects; however, by definition, the neurocrine, like the endocrine, in-volves transport of the hormonal signal via the circulatory system from a point of origin (secretion) in the nerve to a target cell (or tissue) some distance away. (REF. 3, p. 820)

580. C. Depending on the route of transmission, the same mes-senger molecule may function as a neurotransmitter only (axonal transmission), an endocrine hormone only (bloodstream), or both (transmitted both axonally and by the blood as a neurohormone), or the messenger molecule may act locally (paracrine or auto-crine). (REF. 3, p. 820)

581. E. Posttranscriptional modification takes place in *the nu-cleus.* The mRNA directs synthesis of preprohormone which is then cleaved of its N-terminal to form prohormone. The final cleavage takes place in secretory granules and copeptides are syn-thesized and secreted along with the hormone. (REF. 3, p. 821)

582. C. Catecholamines and protein hormones are typical of a unicellular mode of hormone synthesis and release. The hormones are active at the time of release and generally need no further modification to express their activity. Stored in granules, they are modified and released by exocytosis. Another example of a unicel-lular mode of hormone synthesis and release is the production and release of thyroid and steroid hormones. These are simply com-partmentalized during modification and are released into the cy-tosol and subsequently diffuse through cell membrane. (REF. 3, pp. 822–823)

583. D. Vitamin D, a sterol, is synthesized in the skin with very low activity which requires modification by the liver and kidneys to produce its most potent active form. Angiotensin, a peptide hormone, is synthesized from a globulin precursor released from the liver. This globulin precursor is enzymatically and sequentially modified in the kidneys and lungs. Adrenalin is typical of unicellular synthesis and release. Estrogens are produced from androgens in the gonads in a cell-to-cell sequential synthesis pathway. (REF. 3, p. 823)

584. E. All of the factors mentioned in the question are mechanisms associated with hormonal control. Their pattern of hormonal control seems to be dictated by rhythms that may be genetically encoded, such as the patterns of hormone secretion that occur when an individual enters puberty. (REF. 3, p. 824)

585. E. The principle of negative feedback acts to limit output of both hormone and product. In general, anything that changes the level of product serum levels will have a feedback effect on hormone secretion. For example, if a hormone is designed to increase production of product or retard its utilization, it will be inhibited by increased product serum levels. On the other hand, if the secretion of a hormone is designed to retard product production or accelerate product uptake, then the inhibitory effect is a low serum level of product. (REF. 3, p. 824)

586. E. Steroid hormone receptor sites are located in the cytosol and the nucleus. Thyroid hormone receptors are principally in the nucleus, while peptide, protein, and catecholamine hormone receptors are located on the cell surface. These are typically polar moieties which do not cross cell membranes, unlike relatively nonpolar steroid hormones. Hormone receptors are very large protein molecules present in very large numbers, perhaps to avoid problems associated with saturation kinetics. In addition, plasma membrane receptors usually contain carbohydrates and occasionally phospholipids. (REF. 3, p. 824)

587. B. Since the association of a hormone with its receptor is a reversible reaction, the chemical kinetics can be expressed in terms of a K value for the association. Accordingly, this reaction follows saturation kinetics. As the concentration of free hormone increases, $[H] \rightarrow$ infinity, the concentration of bound hormone increases until it reaches receptor capacity, $[HR] \rightarrow$ receptor capacity. At this point all receptors are occupied by hormone. At the same time the ratio of bound hormone to free hormone decreases and approaches zero, $[HR]/[H] \rightarrow 0$. The ratio of bound hormone to free hormone can be plotted as a function of bound hormone. Such plot usually yields a straight line, the slope of which equals the negative of the association constant, and the X intercept equals the receptor capacity. (REF. 3, pp. 825–826)

588. A. As long as the intracellular steps in hormone action are not rate limiting, anything that increases receptor site number or affinity increases the sensitivity of the cell. Factors affecting affinity include: pH, osmolality, and ion concentration, as well as hormone levels. Generally, receptor capacity is regulated by its own hormone. High hormone levels result in down regulation by the cells decreasing the number of receptor sites. Conversely, low hormone levels tend to increase receptor site numbers. (REF. 3, pp. 826–827)

589. D. Lysozymal degradation of the internalized receptor-hormone complex is part of the receptor recycling system of the cell. It is unclear whether or not the internalized complex initiates any intracellular hormonal action before it is disrupted and degraded. (REF. 3, p. 827)

590. B. Binding of hormone to receptor results in conformational changes in the regulatory subunit of adenylate cyclase. This allows binding of GTP, this GTP complex activates the catalytic subunit to act on Mg-ATP to form cAMP. GTP is simultaneously converted to GDP which is then less able to activate the catalytic subunit. (REF. 3, p. 827)

591. A. Although intracellular concentrations of calcium ions may increase, it is the binding of the calcium ion to its binding protein calmodulin in various proportions which acts to amplify cellular enzyme and metabolic activities. The effect of the calcium-calmodulin is to regulate by activation and deactivation of enzymatic and metabolic pathways. (REF. 3, p. 828)

592. B. Steroid hormones, particularly thyroid hormones, are specific for cytoplasmic receptors and induce cellular response by interacting with DNA. This is accomplished by first binding to receptor protein inside the cell. This hormone-receptor complex then enters the nucleus where it combines with acceptor proteins. This combination of hormone receptor/protein receptor interacts with chromatin containing target DNA promoter sites. This process logically explains the usually slower response elicited by steroid hormones. (REF. 3, p. 831)

593. C. The binding of hormone to receptor causes the formation and release of phosphotidyl inositol-4,5-*bis*-phosphate. Cleavage by phospholipase A_2 and C to inositol triphosphate and diacylglycerate (DAG). DAG is cleaved to give arachidonic acid, which is a substrate for prostaglandin synthesis. (REF. 3, pp. 828–829)

594. E. Basal level of activity can be observed independent of added hormone; therefore, there is a minimal threshold of hormone concentration required to elicit a measurable response. Other features include duration of response and the effects of antagonistic or synergistic hormones. (REF. 3, p. 831)

595. E. Hormonal responses are not all-or-none: a minimal concentration is required for a measurable response. The concentration of hormone required to elicit a half-measurable response defines the sensitivity of the target cell. Saturation dose defines the maximal response of the cell. Intrinsic basal level activity can be observed long after any previous exposure and independent of any added hormone. (REF. 3, p. 831)

596. A. Plasma half-life is correlated with extent of protein binding. Thyroid and steroid hormones are generally protein bound which extends the half-life. In some instances, molecules may return to plasma by way of the lymphatic system. (REF. 3, p. 833)

597. B. Metabolic clearance rate is defined as plasma cleared per unit time or mass hormone removed per unit time divided by the circulating mass per unit volume of plasma. MCR = mg/min removed/mg/mL of plasma = mL cleared/min. The ratio of MCR/volume of distribution of hormone = fractional turnover rate (K). (REF. 3, p. 833)

598. C. Under ordinary conditions fat provides 57% of our calories. Carbohydrates provide approximately 43%. The basal metabolic requirements can be expressed in terms of the amount of CO_2 produced per milliliter of O_2 consumed. This is referred to as the respiratory quotient (RQ). Carbohydrates account for only 1%

of the energy stores as compared to protein (23%) and fat (76%). (REF. 3, p. 839)

599. A. The sources of glucose in overnight fasting are gluconeogenesis (25%) and glycogenolysis (75%), with 45% of glucose being used by the brain. (REF. 3, p. 842)

600. A. Leucine is the only amino acid that is not glucogenic. (REF. 3, p. 84)

601. D. Chylomicrons are formed from dietary fat and are absorbed through the intestine. They have a half-life of 5 minutes. They are rapidly acted upon by lipoprotein lipase to liberate free fatty acids in adipose, heart, and muscle tissue. The chylomicron remnants give rise to LDL and bile salts in the liver. (REF. 3, p. 847)

602. E. All hormones listed are correct. (REF. 3, p. 849)

603. D. Insulin:beta cells; glucagon:alpha cells; somatostatin:sigma cells. (REF. 2, p. 849)

604. C. Some stimulators besides carbohydrates include: protein, free fatty acids, ketoacids, glucagon, potassium, calcium, and sulfonylurea drugs. Inhibitors include: fasting, exercise, somatostatin, prostaglandins, and drugs such as diazoxide and phenytoin. (REF. 3, p. 855)

605. C. Very little insulin is excreted unchanged into the urine. It circulates unbound to any carrier proteins. As a result, insulin has a very short half-life, 5 to 8 minutes. Degradation takes place principally in the liver and kidneys. (REF. 3, pp. 857–858)

606. C. Insulin stimulates cellular uptake of amino acids as well as certain ions such as magnesium, potassium, and phosphate. (REF. 3, p. 859)

607. B. Antidiuretic hormone (ADH) is secreted by the posterior pituitary. (REF. 1, p. 877)

608. E. Because of powerful esterases in the blood, acetylcholine has an extremely short half-life. (REF. 1, p. 688 ff)

609. B. Both androsterone and dehydroepiandrosterone are 17-ketosteroids. These are excreted either into the gut in the bile or into the urine. The rate of excretion of 17-ketosteroids in the urine is an index of the rate of androgen production in the body. (REF. 1, p. 961 ff)

610. C. Parathyroid hormone attempts to raise serum calcium if it falls below normal. (REF. 1, p. 948)

611. D. Androgens are not significant for the development of lactation. (REF. 1, p. 994 ff)

612. D. Chorionic gonadotropin rises precipitously during the first trimester of pregnancy. (REF. 1, p. 989)

613. B. Ovulation occurs regularly 14 days before the onset of menstruation. (REF. 1, p. 970)

614. E. Large areas of the anterior hypothalamus, including especially the preoptic area, are concerned with regulation of body temperature. (REF. 1, p. 853 ff)

615. A. Coincidentally with the development of the trophoblast, cells from early fertilized ovum, the hormone chorionic gonadotropin is secreted by the syncytial trophoblastic cells into the fluid of the mother. The rate of secretion rises rapidly to reach a maximum approximately 7 weeks after ovulation and decreases to a relatively low value 16 weeks after ovulation. Without this secretion the uterus (with its implanted ovum) would slough off. (REF. 1, p. 986)

616. D. Testosterone is secreted from the cells of Leydig and reaches peak levels in the late teens and early twenties. (REF. 1, p. 960)

617. C. The primary stimulus for insulin secretion is an increase in blood glucose levels. (REF. 1, p. 929)

618. B. Glucagon stimulates the breakdown of glycogen. (REF. 1, p. 981)

619. B. Glucagon and epinephrine are additive in many of the metabolic effects. (REF. 1, p. 931)

620. B. Glucagon has a significant effect on increasing blood glucose even after all the glycogen in the liver has been exhausted. This results from increasing the rate of gluconeogenesis in the liver cells by increasing extraction of amino acids from the blood, which are then converted to glucose. (REF. 1, p. 931)

621. C. In normal, nonpregnant females both estrogens and androgens are secreted by the ovaries but only estrogens are exclusively secreted in major quantities only by the ovaries. (REF. 1, p. 968)

622. C. In addition to testosterone, small amounts of estrogens are found in the male testes. (REF. 1, p. 961)

623. B. The adrenal cortex secretes at least five different androgens. (REF. 1, p. 919 ff)

624. C. Testosterone production increases rapidly at the onset of puberty and lasts throughout the remainder of life. At puberty estrogens increase 20-fold. Estrogens are produced in subcritical quantities for a short time after menopause. (REF. 1, pp. 964, 978 ff)

625. C. One of the most important male characteristics is the development of increasing musculature following puberty. The bones grow considerably in thickness and also deposit considerable calcium salts. Estrogens cause increased osteoblastic activity and also a slight increase in total body protein. (REF. 1, p. 964)

8 Gastrointestinal Physiology

DIRECTIONS (Questions 626–676): For each of the questions or incomplete statements below, **one** or **more** of the answers or completions given is correct. Select

  A if only *1, 2, and 3* are correct
  B if only *1 and 3* are correct
  C if only *2 and 4* are correct
  D if only *4* is correct
  E if all are correct

626. Amino acid absorption
    1. is stereospecific
    2. does not depend on glucose being present in the duodenal lumen
    3. is linked to Na^+
    4. depends on the Ca^+ concentration in the duodenal lumen

627. When an obese but otherwise normal person goes on a diet deficient in calories
    1. body fat is lost at a rate proportionate to the caloric deficiency of the diet
    2. initial weight loss exceeds initial fat loss
    3. retention of body water obscures the initial fat loss
    4. initial weight loss is due primarily to water loss

Directions Summarized				
A	B	C	D	E
1,2,3	1,3	2,4	4	All are
only	only	only	only	correct

628. The symptoms associated with the congenital absence of lactase
   1. is most frequently observed in infants
   2. is characterized by an intolerance of milk
   3. results in diarrhea following ingestion of lactose
   4. may be present in a person who tolerates sucrose and maltose

629. Gastric secretion normally occurs in response to
   1. sight and smell of food
   2. duodenal distension
   3. emotional disturbance
   4. gastrin release

630. Secretions of the small intestinal mucosa
   1. contain all of the enzymes necessary for digestion
   2. contain disaccharidases that are very important for normal carbohydrate digestion
   3. are stimulated by gastrin
   4. serve primarily a protective function

631. After free fatty acids and glycerol enter the epithelial cells, these compounds
   1. move out of the cell by simple diffusion
   2. are transported out of the cell by specific carrier systems
   3. are mostly carried away by the bloodstream
   4. are resynthesized into triglycerides

632. Which of the following is associated with oxyntic gland secretion
   1. mucus
   2. hydrochloric acid
   3. intrinsic factor
   4. pepsinogen

633. Lipid absorption
1. is facilitated by specific transport proteins in the brush border
2. is facilitated by liver bile
3. depends on the lipid solubility of lipase for hydrolysis of triglycerides
4. is accelerated by the formation of micelles

634. Bile salts
1. are absorbed in the stomach
2. are not reabsorbed
3. are absorbed in the duodenum
4. are absorbed chiefly in the ileum

635. Sodium transport in the GI system
1. is down an electrochemical gradient as it enters the epithelial cells of the duodenum
2. is an active pumping process as it leaves the epithelial cells of the duodenum
3. depends on sodium ion carriers found on the basal and lateral borders of the epithelial cells in the duodenum
4. may include potassium or hydrogen exchange

636. Glucose
1. is linked to sodium ion absorption
2. probably depends on a carrier that requires both sodium ion and glucose for movement of the sugar into the epithelial cell
3. probably requires energy to remove sodium ion from the epithelial cell
4. probably does not require energy for the movement of glucose into the epithelial cell

637. Secretin
1. is secreted by the pancreas
2. increases pancreatic secretion of HCO_3
3. stimulates pancreatic secretion of lysine
4. is released from the mucosa of the small intestine

		Directions Summarized		
A	**B**	**C**	**D**	**E**
1,2,3	1,3	2,4	4	All are
only	only	only	only	correct

638. Vitamin B_{12} absorption
   1. occurs in the stomach
   2. depends on the presence of intrinsic factor
   3. depends on passive diffusion
   4. occurs in the ileum

639. The movement of lipid out of the epithelial cells
   1. depends on the formation of chylomicra
   2. depends on packaging of the lipid in a protein envelope
   3. depends on the lymph system
   4. is abnormal in a beta-lipoprotein

640. Lipid absorption
   1. requires formation of chylomicra
   2. is by diffusion through the brush border
   3. is Na^+ linked
   4. depends on liver bile for a normal time course

641. Absorption of materials from the GI tract
   1. is facilitated by the increase in surface area provided by the physical arrangement of the small intestine
   2. will depend on the ability of compounds with molecular weights over 180 to diffuse through a lipid barrier unless a specific transport system is available
   3. can occur against an electrochemical energy gradient
   4. depends on the active pumping of water

642. Motility of the small intestine is stimulated by
   1. acetylcholine (ACh)
   2. gastrin
   3. cholecystokinin
   4. pancreozymin

643. The act of swallowing is associated with
   1. upper esophageal sphincter relaxation when food is placed in contact with the anterior pillars of the pharynx
   2. concurrent inhibition of respiration
   3. closure of the glottis
   4. movement of food into the nasopharynx

644. Amino acid absorption
   1. may be mediated through a carrier that requires amino acid and sodium ion for activation
   2. may not require energy for movement of amino acid into epithelial cells
   3. may require energy to remove sodium ions from epithelial cells
   4. may use the concentration gradient of sodium ions from the gut lumen to the epithelial intracellular compartment to supply the energy for amino acid movement out of the gut and into the epithelial cells

645. Calcium absorption is enhanced by
   1. high-fat, low-protein diet
   2. acidification of intestinal contents
   3. elevated phosphate intake
   4. vitamin D in the diet

646. The effect of increased vagal stimulation on the intestine is
   1. increased rate of muscle spike potentials
   2. increased oxygen consumption
   3. increased motility
   4. hyperpolarization of smooth muscle membrane

647. Which of the following is associated with the esophageal phase of swallowing?
   1. Relaxation of the upper esophageal sphincter
   2. Sealing off of the nasopharynx
   3. Initiation of the primary phase of peristalsis
   4. Begins when food bolus comes into contact with back of tongue

648. The presentation of a bolus of solid food to the mouth
1. is usually followed by mastication
2. stimulates taste buds
3. reflexively stimulates the salivary glands
4. results in muscular activities that are under conscious control

649. The gallbladder
1. stores bile secreted continuously by the liver
2. can be caused to contract by cholecystokinin
3. can be caused to contract by vagal activity
4. can have its contraction blocked by activity in certain sympathetic nerves

650. Gastric peristalsis
1. originates in the proximal half of the stomach
2. increases in intensity as it sweeps toward the pylorus
3. is characterized by strong contractions of the antrum at the end of the wave
4. ejects all the contents of the antrum into the duodenum

651. Motility of the small intestine is
1. totally independent of activity of the stomach
2. stimulated by adrenergic agents
3. inhibited by the vagus nerve
4. stimulated by radial stretch of the gut

652. Deglutition (swallowing)
1. is an automatic function of smooth muscle
2. is associated with a raising of the soft palate to prevent reflux of food into the nasopharynx
3. does not require relaxation of cricopharyngeal muscle
4. is a complicated act requiring the precise coordination of many muscle groups

653. Esophageal peristalsis
1. is initiated by vagal reflexes
2. can be caused by distension of the esophagus
3. is stimulated by ACh
4. is characterized by activity proximal to the pharynx inhibiting activity distal to the pharynx

654. Defecation
1. is a reflex interruption of anal continence
2. depends on information from stretch receptors in the wall of the rectum
3. can be delayed by conscious contraction of the external sphincter and levator ani muscles
4. requires intact parasympathetic innervation of the rectum for proper reflex integration

655. Liver bile
1. formation requires oxidative metabolism
2. can be secreted against a pressure greater than aortic blood pressure
3. flow from the liver is best stimulated by the presentation of increased amounts of bile to liver cells
4. is secreted by a few select hepatic cells

656. Carbohydrate absorption
1. proceeds mainly with polysaccharides
2. depends on diffusion of glucose through the epithelial cell membranes
3. does not require energy
4. is not the same for fructose and glucose

657. Secretin
1. primarily stimulates enzyme secretion of the pancreas
2. is released from the gastric mucosa
3. release is stimulated by protein hydrolysates in the lumen of the small intestine
4. has no relationship to gastrin

Directions Summarized				
A	**B**	**C**	**D**	**E**
1,2,3	1,3	2,4	4	All are
only	only	only	only	correct

658. Cholecystokinin
1. is released from intestinal mucosa cells
2. release is stimulated by distension of the stomach
3. release is stimulated by protein hydrolysates in the lumen of the small intestine
4. has no structural relationship to gastrin

659. Liver bile flow is increased by
1. pancreatic secretion
2. secretin
3. gastrin
4. vagal stimulation

660. Gastrin
1. increases HCl secretion
2. is released by ethanol in the stomach
3. is released by stomach distension
4. is released by vagal stimulation

661. Motility of the colon is
1. stimulated by ACh
2. primarily inhibited by the vagus nerve
3. inhibited by norepinephrine
4. stimulated by distension of the ileum

662. Salivary secretion
1. comes mostly from the parotid, submaxillary, and sublingual glands
2. has a mucous component
3. has a serous component
4. is largely under hormonal control

663. Hydrochloric acid secretion
1. is accomplished by passive diffusion
2. requires the dissociation of water with subsequent exchange of the hydrogen ion for potassium ion
3. utilizes protein molecules to neutralize OH^- remaining in the secretory cell
4. requires oxidative metabolism

664. The chyme entering the small intestine causes a release of secretin which results in
1. pancreatic fluid secretion of a large volume containing low chloride but high bicarbonate concentration
2. stimulation of pancreatic fluid in which there are no enzymes
3. a pancreatic secretion whose pH is just right for action of the pancreatic enzymes that are eventually released
4. a pancreatic fluid that aids in protection against the development of duodenal ulcers

665. The rate of gastric emptying is
1. affected by the resistance to flow offered by the pylorus
2. increased by the presence of fats, acid, or protein hydrolysates in the duodenum
3. approximately proportional to the volume introduced into the stomach
4. primarily controlled by the glucose concentration in the saliva

666. Vomiting
1. is the forceful expulsion of the contents of the digestive tract through the mouth
2. results in dehydration, depletion of a body HCO_3, Na^+ and K^+, and alkalosis
3. results in loss of fluid, and if prolonged, can result in circulatory collapse and death
4. is a complex reflex act, which is coordinated by a center located in the sacral region of the spinal cord

		Directions Summarized		
A	**B**	**C**	**D**	**E**
1,2,3	1,3	2,4	4	All are
only	only	only	only	correct

667. Oxyntic or parietal cells secrete
   1. zymogen granules
   2. intrinsic factor
   3. pepsinogen
   4. HCl

668. The stomach is a poor area for absorption primarily because
   1. most foods are swallowed before ptyalin has a chance to break down starch
   2. the junction between epithelial cells are tight junctions
   3. pH of the stomach is too high
   4. the stomach lacks villus membranes

669. Absorption of water through the intestinal membrane
   1. occurs when the chyme is dilute
   2. follows usual laws of osmosis
   3. is by simple diffusion
   4. may be transported from plasma into chyme

670. In carbohydrate absorption
   1. transport is selective
   2. it is highest for galactose
   3. carbohydrates are absorbed almost entirely as monosaccharides
   4. carbohydrates display competitive absorption

671. Various secretions along the alimentary tract
   1. serve as lubrication to aid in digestion
   2. are formed in response to ingestion of food
   3. are regulated according to the amount of food consumed
   4. provide protection of the mucosa

672. Stimulation of the gastrointestinal secretions includes:
   1. tactile stimulation
   2. chemical stimuli
   3. distension
   4. irritation

673. The pangs associated with hunger
   1. diminish after 3 or 4 days of starvation
   2. usually appear 12 to 24 hours after fasting begins
   3. are accompanied with feelings of hunger and pain in the pit of the stomach
   4. are increased by a low level of blood sugar

674. Swallowing is dependent upon
   1. ninth cranial nerves
   2. trigeminal nerve
   3. vagus nerve
   4. pyramidal tract

675. The musculature of the esophagus below the pharyngeal junction is
   1. smooth only in lower third
   2. is striated in the upper third
   3. mixed smooth and striated muscle in the middle third
   4. innervated primarily from vagal and glossopharyngeal nerves

676. Enzymes associated with protein digestion include:
   1. pepsin
   2. trypsin
   3. chymotrypsin
   4. carboxypolypeptidase

DIRECTIONS (Questions 677–692): Each of the questions or incomplete statements below is followed by five suggested answers or completions. Select the **one** that is **best** in each case.

677. The sugars normally found in significant amounts in the intestinal chyme include:
    A. glucose and fructose
    B. galactose and xylose
    C. mannose and ribose
    D. mannose and xylose
    E. ribose and xylose

678. Man is unable to digest dietary
    A. glycogen
    B. dextrin
    C. saccharose
    D. cellulose
    E. glucose

679. Specific dynamic action is
    A. the specific effects of pharmacologic agents on the GI tract
    B. the effect of certain foodstuffs on metabolic rate
    C. the effect of parasympathetic stimulation on metabolic rate
    D. the effect of sympathetic stimulation on metabolic rate
    E. the effect of exercise on metabolic rate

680. Gastric secretion is
    A. increased by stomach distension
    B. stimulated by an increase in phonic activity
    C. stimulated by norepinephrine
    D. inhibited by curare
    E. not affected by the presence of food in the stomach

681. Which of the following is NOT associated with pancreatic secretions?
- **A.** Source of HCO_3 for the neutralization of gastric acid in the small intestine
- **B.** Contains most of the digestive enzymes
- **C.** Reduces the osmolality of the fluid in the gut
- **D.** Source of tributyrase
- **E.** Has a pH that is primarily controlled by ventilation of the lungs

682. Primary peristalsis and secondary peristalsis of the esophagus are different in that the latter
- **A.** is more rapid
- **B.** is independent of neural control
- **C.** is initiated by swallowing
- **D.** is confined to the upper esophagus
- **E.** is under intrinsic neuronal control

683. Which of the following is NOT associated with the motor function of the stomach?
- **A.** Storage
- **B.** Mixing
- **C.** Chyme formation
- **D.** Rapid emptying to accommodate entry of excess food into the stomach
- **E.** None of the above are associated with the motor function of the stomach

684. In Figure 22, the major pathway(s) for water movement from the mucosal side to the serosal side of the intestinal mucosa is(are) indicated by which arrows?
- **A.** 1, 2, and 3
- **B.** 1 and 3
- **C.** 2 and 4
- **D.** 4 only
- **E.** All of the above

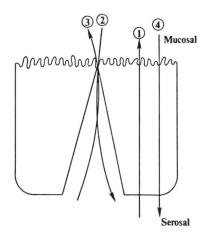

Figure 22.

685. The secretion of intrinsic factor occurs in the
    A. parietal cells of the stomach
    B. chief cells of the stomach
    C. upper duodenum
    D. beta cells of the pancreas
    E. liver

686. Ingested cholesterol enters the intestinal epithelial cell by
    A. diffusion through the lipid portion of the cell membrane
    B. diffusion through pores in the cell membrane
    C. an active transport mechanism
    D. pinocytosis
    E. esterification, diffusion through pores, and then hydrolysis

687. Vitamin D is essential for normal
    A. glucose absorption
    B. fat absorption
    C. protein absorption
    D. Ca^{2+} absorption
    E. antidiuretic hormone (ADH) secretion

688. Salivary secretion
   A. has a constant composition regardless of the rate of secretion
   B. is probably a simple ultrafiltrate of plasma
   C. has the same constituency no matter what kind of material is placed in the mouth
   D. is entirely under neural control
   E. is entirely under cortical control

689. The pacemaker for small intestine motility
   A. lies in the fundal region of the stomach
   B. controls the motility of all of the small intestine
   C. only controls a small area of gut near the bile duct
   D. lies in the radial muscle of the gut near the pyloric sphincter
   E. lies in the cerebrum (pacemaker cells are not present in the GI system)

690. Distension of the stomach
   A. is associated with a decrease in peristaltic activity in the stomach
   B. decreases tone of the lower esophageal sphincter
   C. causes an acute increase in pressure inside the resting stomach
   D. results in a potentially large increase in volume with very little pressure change
   E. is under cerebral control, as it is solely influenced by psychic factors

691. Secretin is released by
   A. acid in the duodenum
   B. acid in the urine
   C. S cells in the duodenal mucosa
   D. distention of the colon
   E. cells in the hypothalamus

692. The secretion of bile from the liver
   A. is not necessary for normal digestion
   B. contains large amounts of high-protein solution
   C. is important only for the normal digestion of proteins
   D. is stored in the gallbladder
   E. is important only for the normal digestion of sugars

DIRECTIONS (Questions 693–696): The group of questions below consists of a set of lettered components followed by a list of numbered words or phrases. For **each** numbered word or phrase, select the **one** lettered component that is most closely associated with it. Each lettered component may be selected once, more than once, or not at all

---

   A. Protein decarboxylase
   B. Vitamin A
   C. Pantothenic acid
   D. Folinic acid
   E. Riboflavin

693. Required with thiamine in decarboxylation of pyruvic acid

694. Required as part of the coenzyme A molecule

695. Definite nutritional requirement

696. Required for incorporation into hydrogen carrier coenzymes

DIRECTIONS (Questions 697–700): The set of lettered items below is followed by a list of numbered words or phrases. For each numbered word or phrase select

A if the item is associated with **A** only
B if the item is associated with **B** only
C if the item is associated with both **A** and **B**
D if the item is associated with neither **A** nor **B**

    A. Sympathetic activity
    B. Parasympathetic activity
    C. Both
    D. Neither

697. Required for gastric phase of digestion

698. Required for cephalic phase of digestion

699. Required for mucous secretion of the large intestine

700. Involved in mucous secretion of the salivary glands

Explanatory Answers

626. A. Amino acid absorption in the gut is stereospecific, is linked to sodium movement, and does not require glucose. (REF. 1, p. 790)

627. B. When an obese individual starts a calorie-deficient diet, fat is utilized immediately to make up for the missing calories. In addition to this fat utilization, there is a retention of body fluid. (REF. 1, p. 788 ff)

628. E. Infants with congenital absence of lactase frequently have a watery diarrhea, but these same infants may tolerate sucrose and maltose perfectly well. (REF. 2, p. 1454 ff)

629. D. Gastrin does not require an intact reflex pathway to elicit increased gastric secretion. (REF. 3, p. 698)

630. D. The major secretions of the small intestine are mucous secretions for the protection of the intestinal lining. (REF. 1, p. 784)

631. D. After free fatty acids and glycerol enter the epithelial cells of the gut they are reesterified to form triglycerides before they leave the intracellular pool. (REF. 1, p. 795)

632. E. Mucus, hydrochloric acid, intrinsic factor, and pepsinogen are secretions associated with the oxyntic gland found in the stomach. (REF. 1, p. 774 ff)

633. C. The formation of micellar aggregates by liver bile is crucial for the normal absorption of lipids. (REF. 1, p. 788 ff)

634. D. Bile salts are largely absorbed by an active transport system in the ileum. (REF. 1, p. 788)

635. E. Sodium ion transport is down an electrochemical gradient as it enters cells of the gut but requires active transport to leave the cells of the gut. The active transport is found on the basal and lateral borders of the epithelial cells in the duodenum and is stimulated by glucose. (REF. 1, p. 792 ff)

636. E. Glucose absorption uses the normal extracellular/intracellular sodium ion gradient to assist the movement of the sugar into the epithelial cells of the gut. The sodium ion must then be removed by an energy utilizing Na^+/K^+ ATPase. (REF. 1, p. 794)

637. C. The hormone secretin is secreted by the small intestine. It causes increases in fluid and bicarbonate secretion by the pancreas. (REF. 1, p. 780)

638. C. The absorption of vitamin B_{12} by the ileum requires intrinsic factor from the stomach. (REF. 1, p. 869)

639. E. Once lipid material has moved into epithelial cells of the gut it is packaged in protein envelopes to form chylomicra, which are then dumped into the lymphatic system. In a beta-lipoproteinemia, this system is interrupted because adequate envelope protein is not available for producing chylomicra. (REF. 1, p. 795)

640. C. Liver bile assists the diffusion of lipid digestive products across the brush border of the ileum by emulsifying the fats to increase their surface area. (REF. 1, p. 788)

641. A. Absorption from the lumen of the gut is limited by the lipid solubility of compounds with molecular weights over 180 and can occur against an electrochemical energy gradient in cases of active transport. To facilitate this absorption, the gut is folded to increase the surface area available for movement of materials. (REF. 3, p. 763)

642. E. Motility of the ileum is stimulated by gastrin, cholecystokinin-pancreozymin, and by ACh released from parasympathetic innervation. (REF. 3, p. 764)

643. A. When swallowing occurs the shift from the more common movement of air in and out of the lungs to the movement of solids or liquids into the stomach requires inhibition of respiration, closure of the glottis, and relaxation of the upper esophageal sphincter. (REF. 2, p. 1426 ff)

644. E. Amino acid absorption uses the sodium ion gradient from the lumen into the intracellular fluid of the gut epithelium to drive amino acid transport. The sodium then must be removed from the epithelial cells by active transport. (REF. 1, p. 1454 ff)

645. C. A low intestinal pH tends to keep calcium in solution and, thus, favors absorption. Vitamin D enhances the absorption of Ca^{2+} from the gut. (REF. 1, p. 793)

646. A. Increased vagal stimulation depolarizes smooth muscle cells. (REF. 1, pp. 141, 757)

647. A. Relaxation of the upper esophageal sphincter with the subsequent initiation of the primary wave of peristalsis is associated with phase III or the esophageal phase. (REF. 2, p. 1426 ff)

648. E. When a bolus of food is presented to the mouth, voluntary activities of mastication ensue with taste bud activation and reflex stimulation of the salivary glands. (REF. 2, p. 1426 ff)

649. E. The gallbladder is the primary storage site of liver bile and can be stimulated to contract by cholecystokinin and vagal activity. Sympathetic activity can be antagonistic. (REF. 1, p. 781 ff)

650. A. Gastric peristalsis is a wave of muscular contraction which starts in the fundus of the stomach and sweeps toward the antrum with a constantly increasing force of contraction. (REF. 1, p. 758)

651. D. The primary intrinsic stimulus for the contraction of the smooth muscle of the ileum is radial stretch. (REF. 1, p. 766)

652. C. Deglutition is a complicated act requiring precise coordination of many structures. A small portion of this activity is the raising of the soft palate to prevent reflux of food into the nasopharynx. (REF. 1, p. 760)

653. E. Esophageal peristalsis is a wave of muscular contraction along the esophagus that closes the lower esophageal sphincter, can be stimulated by radial stretch of the esophagus, is stimulated by ACh, and is characterized by activity proximal to the pharynx inhibiting activity distal to the pharynx. (REF. 1, p. 761 ff)

654. E. Defecation is a reflex activity using stretch receptors in the rectum and the parasympathetic nervous system in its reflex arc. This reflex can be overcome by the conscious contraction of the external sphincter. (REF. 1, p. 768)

655. A. Liver bile formation is an active process requiring energy. This energy is evident when it is noted that bile can be secreted against a pressure greater than aortic blood pressure. One of the best stimulants for bile secretion is the presentation of small amounts of bile to the blood perfusing the liver. (REF. 1, p. 781 ff)

656. D. Carbohydrate absorption is an active carrier-mediated process which acts only on monosaccharides. The carriers for fructose and glucose appear to be separate systems. (REF. 1, p. 794)

657. D. The major action of the hormone secretin is to stimulate bicarbonate and fluid secretions by the pancreas. (REF. 2, p. 1450 ff)

658. A. The release of pancreozymin (now called cholecysto-kinin) from intestinal mucosa cells is stimulated by distension of the stomach and the presence of protein hydrolysates in the lumen of the small intestine. The amino acid sequence of gastrin and pancreozymin do have some elements in common. (REF. 2, p. 1449 ff)

659. C. Secretin and vagal stimulation can increase the flow of liver bile. (REF. 1, p. 781)

660. E. The major function of gastrin is to stimulate the secretion of HCl into the stomach as a response to food being presented to the upper GI tract. The release of gastrin can be stimulated by ethanol in the stomach, stomach distension, and vagal stimulation. (REF. 1, p. 763)

661. B. Motility of the colon is controlled by a dual system of innervation which contains cholinergic nerves that excite and adrenergic nerves that inhibit the colon. (REF. 2, p. 1423 ff)

662. A. Salivary secretions have both a mucous and serous component and are obtained from the carotid, submaxillary, and sublingual glands. The control of these secretions is entirely integrated by the CNS. (REF. 2, p. 1439 ff)

663. C. The secretion of HCl by the gastric mucosa is an active energy-dependent process which involves the dissociation into H^+ and OH^- with active secretion of H^+ in exchange for K^+ and will combine with the actively secreted Cl^+ in the canaliculus. (REF. 1, p. 775)

664. E. Pancreatic secretion is humorally regulated by the hormones secretin and cholecystokinin. The chyme (which contains HCl) causes the small intestine to release and activate the hormone secretin which acts on the pancreas to release a large volume of fluid. The fluid contains a high concentration of bicarbonate (145 mEq/L) but a low concentration of chloride ion. This fluid, however, contains almost no enzymes when the pancreas is stimulated by secretin alone. It is the proteoses and peptones in the chyme that cause the small intestine to secrete a second hormone, cholecystokinin, which is responsible for the secretion of the pancreatic digestive enzyme. The large bicarbonate secretion by the pancreas is important in that it neutralizes the acid entering the duodenum and thus provides a protective effect against duodenal ulcers. (REF. 1, p. 779 ff)

665. B. The rate of gastric emptying is partially determined by the resistance to flow of the pylorus and will increase as more material is placed in the stomach. (REF. 1, p. 763 ff)

666. A. The forceful expulsion of ions and fluid leads to alkalosis. Constant vomiting can be life-threatening. A vomiting center may be located in the medulla. (REF. 1, p. 803)

667. C. Parietal cells are responsible for the secretion of HCl and intrinsic factors into the stomach. (REF. 1, p. 776)

668. C. The barriers are anatomical; i.e., tight junctions in the epithelium and lack of villus membranes. (REF. 3, pp. 790–791)

669. E. The absorption of water through the intestinal membrane follows the usual laws of osmosis, therefore when the chyme is dilute water goes from the gut lumen into the intestinal membranes. The flow is by simple diffusion. (REF. 1, p. 792)

670. E. Galactose has the highest absorption rate for the carbohydrates. The absorption of one carbohydrate tends to decrease absorption for another. Transport is selective and competitive. (REF. 2, p. 1058)

671. E. As indicated in the question. (REF. 1, p. 770)

672. E. Tactile or chemical irritation elicits reflexes that activate the enteric nervous system. Distension can elicit increased motility which in turn promotes secretion. (REF. 3, pp. 770–771)

673. E. These pangs are associated with low blood glucose levels and pain but the cause of hunger pangs is unclear, as is their contribution to the control of food intake. However, they do appear 12 to 24 hours after fasting begins and gradually weaken until they disappear after 3 or 4 days. (REF. 1, p. 763)

674. B. The ninth and tenth cranial nerves are necessary for normal swallowing. (REF. 1, p. 760 ff)

675. E. The vagus nerve supplies the striated muscle which makes up most of the upper esophagus. The middle third of this tube is mixed smooth and striated muscle while the lower third is almost entirely smooth muscle. (REF. 1, p. 761 ff)

676. E. Pepsin is secreted by the stomach, whereas trypsin, chymotrypsin, and carboxypolypeptidase are secreted by the pancreas. (REF. 1, p. 790)

677. A. Glucose and fructose are the results of sucrose hydrolysis and are very prominent in intestinal chyme. (REF. 1, p. 788)

678. D. Man is unable to digest dietary cellulose, because there are no enzymes in the human alimentary tract capable of digesting it. (REF. 3, p. 718)

679. B. Specific dynamic action is the ability of certain amino acids derived from protein foods to generate more heat than their caloric value. (REF. 1, p. 846)

680. A. Gastric secretion is stimulated by stomach distension, vagal activity, and acetylcholine. (REF. 1, p. 774 ff)

681. A. Pancreatic juice has several important actions. It neutralizes the acidic chyme from the stomach, reduces the osmolality of fluid from the gut, and contains most of the major digestive enzymes. Tributyrase, or gastric lipase, is secreted in the stomach and is associated with triglyceride digestion. (REF. 1, p. 778 ff)

682. E. Secondary peristalsis is a local response to distension of the esophagus, while primary peristalsis is initiated by swallowing. Esophageal peristalsis functions to propel food the full length of the esophagus and is under neuronal control. However, secondary peristalsis is under enteric neuronal control. (REF. 1, p. 758 ff)

683. D. Storage and mixing with subsequent chyme formation is followed by slow expulsion. (REF. 1, p. 762 ff)

684. C. Water is absorbed from the lumen of the gut both transcellularly and across the "tight junction" between cells. (REF. 1, p. 742)

685. A. Intrinsic factor for the absorption of vitamin B_{12} is secreted by the parietal cells of the stomach. (REF. 3, p. 695)

686. A. Cholesterol depends of its lipid solubility to diffuse through the cell membrane. (REF. 1, p. 788)

687. D. Vitamin D is necessary for the movement of Ca^{2+} out of the gut lumen into the epithelial cells and out of the epithelial cells into circulation. (REF. 1, p. 870 ff)

688. D. Salivary secretion depends entirely on neural control mechanisms, mainly parasympathetic. The salivatory nuclei (in medulla and pons) are activated by tasting and tactile stimulation from the tongue and mouth. (REF. 1, p. 773 ff)

689. B. The pacemaker for small intestine motility lies in the longitudinal muscle coat near the bile duct. This pacemaker may dictate activity to all of the small intestine. (REF. 2, pp. 218–220, 1431 ff)

690. D. As material enters the stomach, the muscle in the walls will relax to allow a relatively large change in volume with a small pressure change. Part of this relaxation process is due to the property of plasticity inherent in visceral smooth muscle (stress relaxation). (REF. 1, p. 146)

691. C. Secretin is released by cells in the intestinal mucosa (the "S" cells) in response to acid in the duodenum. (REF. 1, p. 780)

692. D. Liver bile, which contains lecithin and cholesterol, is secreted continuously and stored in the gallbladder for periodic release to the gut lumen. (REF. 1, p. 781)

693. A. Protein decarboxylase and thiamine are both necessary for the decarboxylation of pyruvic acid. (REF. 1, p. 867)

694. C. Pantothenic acid is a part of the coenzyme A molecule. (REF. 2, p. 870)

695. B. The fat-soluble vitamin A is required for good vision. (REF. 1, p. 860)

696. E. Riboflavin (Vitamin B_2) is a crucial element in the formation of two coenzymes that operate as hydrogen carriers within the mitochondria. (REF. 1, p. 868)

697. D. The gastric phase of digestion occurs once the food enters the stomach, and it excites the gastrin mechanism and causes local reflexes to become active. (REF. 1, p. 776 ff)

698. B. The cephalic phase of gastric secretion occurs even before food enters the stomach. Impulses from higher centers act upon the dorsal motor nuclei of the vagus which transmit excitation of the stomach. (REF. 1, p. 776 ff)

699. D. The mucous glands of the large intestine are regulated principally by direct tactile stimulation of the cells. (REF. 1, p. 785 ff)

700. A. Sympathetic stimulation of salivary glands causes vaso-constriction and results only in viscous solution to be released from the mucous portion of the glands. (REF. 1, p. 773)

References

1. Guyton AC: *Textbook of Medical Physiology*, 7th ed. W.B. Saunders, Philadelphia, 1986.

2. Patton HD, Fuchs AF, Bertil H, Scher AM, Steiner R: *Textbook of Physiology*, 21st ed. W.B. Saunders, Philadelphia, 1989.

3. Berne RM, Levy MN (eds): *Physiology*, 2nd ed. C.V. Mosby, St. Louis, 1988.